BIOINFORMATICS AND RNA

Innovations in Big Data and Machine Learning

Series Editors

Rashmi Agrawal and Neha Gupta

Artificial Intelligence and Internet of Things: Applications in Smart Healthcare
Lalit Mohan Goyal, Tanzila Saba, Amjad Rehman, and Souad Larabi

Bioinformatics and RNA: A Practice-Based Approach
Dolly Sharma, Shailendra Singh, and Mamta Mittal

For more information about this series, please visit: https://www.routledge.com/Innovations-in-Big-Data-and-Machine-Learning/book-series/CRCIBDML

BIOINFORMATICS AND RNA

A Practice-Based Approach

Dolly Sharma, Shailendra Singh,
and Mamta Mittal

CRC Press
Taylor & Francis Group
Boca Raton London New York

CRC Press is an imprint of the
Taylor & Francis Group, an **informa** business

First edition published 2022
by CRC Press
6000 Broken Sound Parkway NW, Suite 300, Boca Raton, FL 33487-2742

and by CRC Press
2 Park Square, Milton Park, Abingdon, Oxon, OX14 4RN

© 2022 Dolly Sharma, Shailendra Singh, and Mamta Mittal

CRC Press is an imprint of Taylor & Francis Group, LLC

Bioinformatics is everywhere, from disease detection to disease prevention and curation, from the study of microorganisms to study of all life on earth and from sequence alignment to gene expression. This book provides in-depth information regarding the explosively growing new discipline i.e. Bioinformatics. It provides coverage of important topics like sequence alignment, phylogenetics, DNA, RNA, and proteins. From basics of central dogma of biology to computational algorithms for RNA secondary structure prediction, this book covers it all. Also, it provides implementation of few important algorithms in Dot Net framework. This book also enlightens the application areas of bioinformatics.

ISBN: 978-0-367-61909-1 (hbk)
ISBN: 978-0-367-62057-8 (pbk)
ISBN: 978-1-003-10773-6 (ebk)

Typeset in Times
by MPS Limited, Dehradun

Contents

Preface

The book aims to take the reader on a technological voyage of explosively growing new discipline – bioinformatics. After covering the most important topic of this area i.e. alignment, this book further covers phylogenetics. Also, it shelters the less explored topics such as pseudoknot grammar and its types. The book also highlights computational techniques used in bioinformatics to solve various problems. Lastly, it introduces the grey areas of bioinformatics. This book also contains implementation of few important algorithms in Dot Net framework with code.

ORGANIZATION OF THE BOOK

The book is organized so as to include related rudiments and applications of bioinformatics in various areas. The book comprises eight chapters. A brief description of each chapter of this book is as follows.

INTRODUCTION TO BIOINFORMATICS

Chapter 1 first describes the term *bioinformatics*, followed by a debate on the relevance of bioinformatics and its areas. The building blocks of human life, DNA, RNA and protein, are further addressed. Finally, the issue of protein folding, genomics and proteomics is discussed. Elaboration of the central dogma of molecular biology is also done.

COMPUTATIONAL BIOLOGY

Chapter 2 focuses on a very significant aspect of bioinformatics, i.e. the alignment of sequences. Pictorial representations are created step by step in the process of performing global sequence alignment and local sequence alignment. Additionally, multiple sequence alignment is also discussed.

PHYLOGENETICS

Chapter 3 elaborates on phylogenetics as a science of studying biological relationships between organisms. Further, molecular phylogenetics is explained. The latter half of this chapter focuses on phylogenetic tree construction. Using distance matrix methods and character-based methods, the step-by-step process for phylogenetic tree construction is explained.

RNA

Chapter 4 introduces RNA as one of life's most important building blocks, with DNA and protein being the others. RNA is the most versatile agent among these molecules. Various RNA types have been discussed, such as rRNA, tRNA, snRNA, mRNA, etc. Lastly, RNA features are discussed at length.

PSEUDOKNOT

Chapter 5 introduces pseudoknot as an important motif in RNA secondary structure. As pseudoknot is not a real knot, real pictures have been shown to differentiate between them. Various example structures of pseudoknots are shown and various representations of pseudoknots are discussed. Finally, all types of pseudoknots are shown with structures.

PSEUDOKNOT PREDICTION TECHNIQUES

Chapter 6 presents, categorizes and contrasts current prediction models for pseudoknot. These models are divided into six classifications: dynamic programming, comparative approach, heuristic approach, formal grammar, inverse folding, integer programming and inverse folding technique models. The pros and cons of these approaches are discussed.

PSEUDOKNOT GRAMMAR

Chapter 7 begins with an introduction to the Chomsky hierarchy of formal grammars, accompanied by a discussion of the relation between these grammars. The correlation of RNA secondary structure and context free grammar is shown. Recently, formal grammar techniques are used to predict pseudoknots. Various types of grammar are further discussed: Parallel Communicating Grammar System, Pair Stochastic Tree Adjoining Grammar, Context Free Grammar, Multiple Context Free Grammar, Context Sensitive Grammar, Tree Augmented Grammar and Path Controlled Grammar.

NEW AREAS OF BIOINFORMATICS

Chapter 8 illuminates the latest and evolving fields of bioinformatics. Drug target identification and identification of target are discussed. Further, nutrigenomics and toxicogenomics are seen as important areas in bioinformatics that have a significant effect on human health. Finally, bioterrorism is deliberated, which is the hottest topic of today because of the current global pandemic SARS-COVID-19.

Authors

Dolly Sharma, PhD, is an associate professor at Amity University, Noida. She earned an undergraduate degree in computer science and engineering at Kurukshetra University in 2004; a master's in information technology with honors at Panjab University, Chandigarh in 2007 (she was the second University topper); and a PhD in computer science and engineering at Punjab Engineering College, Chandigarh. Dr. Sharma has a rich teaching experience of fourteen years. Her areas of interest include bioinformatics, grid computing and data mining. She has published several research papers and book chapters that are indexed in SCOPUS and SCI. Dr. Sharma has contributed as a reviewer at important conferences and in journals and has chaired international conferences and delivered invited talks. Also, she has filed two patents. Dr. Sharma has supervised many MTech students. She has published one book with Springer and another with Vayu Publishers. Dr. Sharma is a lifetime member of ISTE and AIENG.

Shailendra Singh, PhD, is a professor in the Computer Science and Engineering Department at Punjab Engineering College (Deemed to be University), Chandigarh, India. He earned a BTech in computer science and engineering at HBTI, Kanpur, India; an ME in computer science and engineering at Thapar University, Patiala, India; and a PhD in computer science and engineering at Punjabi University, Patiala, India. His research interest includes bioinformatics, natural language processing, speech technology and soft computing. He has received several awards and recognition in his field and is a member of various professional societies, such as IEEE, IEEE Computational Intelligence Society, Computer Society of India, etc. He has contributed various research papers in international and national journals and conferences.

Mamta Mittal, PhD, is an assistant professor in the Computer Science and Engineering Department at G. B. Pant Government Engineering College, Okhla, New Delhi (under the Government of NCT Delhi). She earned an undergraduate degree in computer science and engineering at Kurukshetra University, Kurukshetra; a master's degree (with honors) in computer science and engineering at YMCA, Faridabad; and a PhD in computer science and engineering at Thapar University, Patiala. Her research area includes data mining, big data and machine learning with rich teaching experience of sixteen years with an emphasis on data mining, machine learning, soft computing and data structure. Dr. Mittal is a lifetime member of CSI and has published and communicated more than 65 several research papers in SCI, SCIE and Scopus indexed journals and attended many workshops, FDPs, and seminars. Two patents published on human surveillance systems and wireless copter for explosive handling and diffusing and one patent granted to her on automated method for tumor detection. She is supervising PhD candidates at GGSIPU (Guru Gobind Singh Indraprastha University), Dwarka, New Delhi. Dr. Mittal is the main editor of Data Intensive Computing Application for Big Data and Big Data Processing Using Spark in and managing editor of International Journal of Sensors, Wireless Communications and Control. She is working on a DST-approved project, "Development of IoT-Based Hybrid Navigation Module for Midsized Autonomous Vehicles", with a research grant of 25 Lakhs and Handing Pradhan Mantri YUVA project for Entrepreneur Cell Activity as faculty coordinator/representative from G. B. Pant Government Engineering College. Dr. Mittal is reviewer of many reputed journals and has chaired several conferences.

1 Introduction to Bioinformatics

1 INTRODUCTION TO BIOINFORMATICS

The term *bioinformatics* is a combination of "bio" + "informatics". Here "Bio" stands for biology and "Informatics" related to information technology. Bioinformatics means informatics techniques are applied to biological problems for finding solutions. The term *bioinformatics* was coined in 1970 by a Dutch Article. Various researchers have defined bioinformatics as

. . . the study of how information is represented and anlyzed in biological systems, information derived at molecular level [1].

Bioinformatics is conceptualizing biology in terms of macromolecules (in the sense of physical-chemistry) and then applying "informatics" techniques (derived from disciplines such as applied maths, computer science, and statistics) to understand and organize the information associated with these molecules, on a large-scale [2].

Bioinformatics has also been stated as a marriage between computer science and molecular biology. Bioinformatics is an inter-disciplinary area of research that integrates biology with information technology, mathematics, statistics, physics and chemistry, as shown in Figure 1.1.

Today is the era of big data. A large number of bioinformatics projects subsist today to handle bioinformatics' big data. The first project of bioinformatics was the Human Genome Project in 1990. The goal of this project was sequencing and annotation of the human genome. Another project, the Celera Genome Project, started in 1998. This project worked on the yeast genome. With the successful completion of these projects, several other projects were initiated; precisely, in 2001, the HapMap Project began.

For carrying out such projects, the need to create, store and communicate huge databases arose. Moreover, biological data was needed in computer-readable form. Consequently, databases like GenBank [3], Ensembl [4], PubMed, M5NR [5], SWISS-PROT [6], OMIM [7], PDB [8], KEGG [9], etc. were coined. Today, large numbers of projects exist for bioinformatics because of the increasing number of researchers in the field of bioinformatics, the explosion of biological data and the increase in the number of funding agencies for projects. There are a number of ongoing projects in various bioinformatics areas like sequence analysis, genomics, structural bioinformatics and computational evolutionary biology. These areas have been illustrated in Figure 1.2.

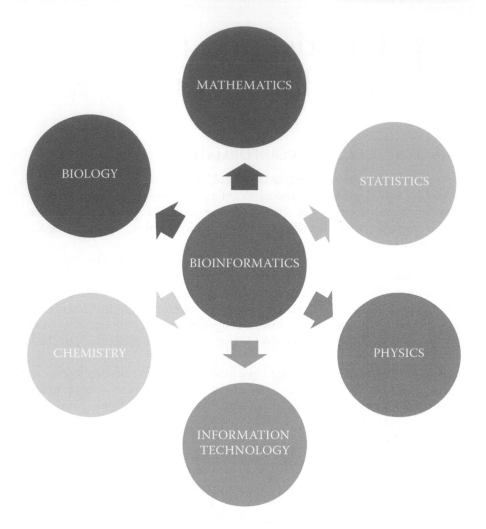

FIGURE 1.1   Bioinformatics: An Inter-disciplinary Area

1.1   IMPORTANCE OF BIOINFORMATICS

Bioinformatics is everywhere, from disease detection to disease prevention and curation, from the study of microorganisms to the study of all life on earth and from sequence alignment to gene expression.

Following are the important areas and applications of bioinformatics:

- Early detection of diseases
- Diagnose and curation of diseases through RNA interference therapeutics
- Discover uncovered facts of living system
- Tools for handling large-scale biological data
- Study of phlogeny of all life on earth

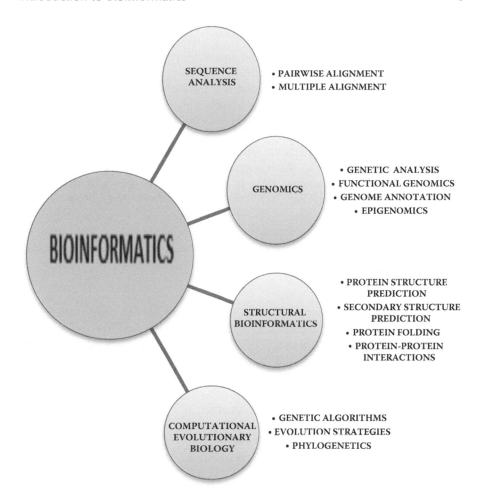

FIGURE 1.2 Bioinformatics Areas

- Prediction of structure of molecules
- Sequence alignment
- Analyzing the function and structure of genomes
- Hybridization of crops
- Expression, interaction and production of proteins
- DNA sequencing for medicine and forensics
- Different aspects of biotechnology including pharmaceutical and micro-biological industries
- Control over diseases through under-expression and over-expression of genes
- Study susceptibility of a person to various diseases through preventive medicine
- Study response of human body of an individual through pharma-co-genomics
- Isolation of genes that enable survival of microorganisms

1.2   DNA, RNA AND PROTEIN SEQUENCES

The house of bioinformatics is based on foundation of DNA, RNA and protein. These three are the building blocks of life.

Organisms have cells that communicate and interact to form tissues, organs and organisms. Cells are made of membranes, proteins, carbohydrates, vitamins and nucleic acids. Bunches of cells form tissue. Lots of tissues form an organ and organs make a whole organism.

Assembly of cell parts requires additional information that is contained in deoxyribonucleic acid (DNA). After assembly, the complete functionality of the body is taken care of by proteins. DNA is converted to proteins with RNA (ribonucleic acid) as an intermediate.

1.2.1   DNA

DNA is present inside the nucleus of cell. It is the cookbook for manufacturing proteins. It carries all the genetic information of a cell. Features of DNA are as follows:

- It is a nucleic acid that encodes information necessary to build a cell.
- It contains hereditary information that may be passed from one generation to another. For example, the colour of allele of the child's eye is the same as that of the mother's eye.
- It stores and maintains cellular information and passes it from generation to generation.
- It is a cookbook for synthesis of proteins.
- DNA is a stable molecule.
- A single copy of DNA exists per cell.
- DNA replicates itself every time the cell is divided.
- DNA structure
  - o   DNA is a double helix structure, as shown in Figure 1.3.
  - o   It contains four bases, namely A for adenine; G for guanine; C for cytosine; and T for thymine.

1.2.2   RNA

DNA is converted to RNA before being transformed to protein. RNA is another nucleic acid similar to DNA is some respects.

The features of RNA are as follows:

- RNA assists DNA in synthesizing a particular protein.
- RNA is located in the nucleus, cytoplasm or ribosome i.e. inside and outside of the nucleus.
- RNA is an unstable molecule.
- Multiple copies of RNA exist per cell.
- There are many types of RNA, namely mRNA, tRNA, rRNA, snRNA, siRNA, miRNA, asRNA
- RNA structure

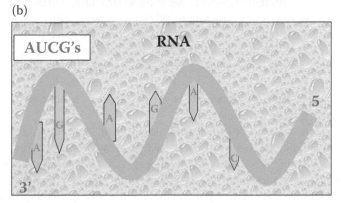

(a)

(b)

FIGURE 1.3 **(a)** DNA Structure; **(b)** RNA Structure

- RNA is a single-stranded structure, as shown in Figure 1.3.
- It contains four bases, namely A for adenine; G for guanine; C for cytosine; and U for Uracil (contrary to thymine in DNA).
- RNA structure has fourlevels – primary structure, secondary structure, tertiary structure and quaternary structure.

1.2.3 Protein

Proteins are the building blocks of life. Proteins are "us". Our body depends on protein for its functioning. There are three types of protein:

- Fibrous Proteins – These proteins form muscles fibre, connective tissue and bone. For example – keratin, actin, myosin, collagen etc.
- Globular Proteins – These proteins carry out functions like transportation, catalyzing and regulation. For example – haemoglobin, globulin, thrombin etc.
- Membrane Proteins – These proteins have multiple roles like connecting cells with each other, receptor proteins and transporting molecules. For example – histones, glucose transporter etc.

They are made up of twenty amino acids that are linked together by bonds called peptide bonds. The source of amino acids in our body is our diet. A few amino acids

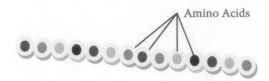

FIGURE 1.4   Protein Primary Sequence

are alanine, lysine, glutamine, threonine and aspartic acid. There are four levels of protein structure – primary structure, secondary structure, tertiary structure and quaternary structure. Primary structure or primary sequence (Figure 1.4) is formed by a sequence of thousands of amino acids. Until this sequence folds, it is termed *primary sequence*. When this amino acid sequence folds and amino acids are linked through hydrogen bonds, the resulting structure is called a secondary structure of protein.

The secondary structure further exists in two forms – α-helix structure and β-pleated sheet – as shown in Figure 1.5.

Further, the secondary structure folds upon itself to form a tertiary protein structure i.e. bonding between α-helices and β-pleated sheets. This is a three-dimensional structure, as shown in Figure 1.6(a). A quaternary protein structure is formed when two or more amino acid chains bond together to form a structure. Amino acid chains are also called polypeptide chains. The quaternary structure of a protein is shown in Figure 1.6(b).

Following are the functions of proteins:

- Transport – Haemoglobin protein transports oxygen in our body.
- Structural Components – Keratin protein forms structures for our skin and nails.
- Hormones – Receptor proteins receive signals and send them further to other proteins.

(a)   (b)

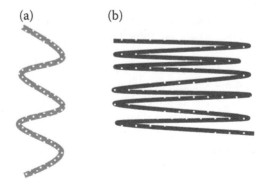

FIGURE 1.5   Protein Secondary Structure: (a) α-Helix Protein Structure; (b) β-Pleated Sheet of Protein

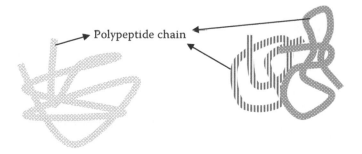

FIGURE 1.6   (a) Tertiary Structure of Protein; (b) Quaternary Structure of Protein

- Contractile Structures – These proteins in our muscles lead to movements in our body; for example, actin and myosin.
- Antibody proteins protect us from disease-causing viruses.
- Enzymes – These proteins catalyze or speed up various biochemical reactions. They are active in our cells all the time.

1.3   CENTRAL DOGMA OF MOLECULAR BIOLOGY

Molecular biology is the biology of molecules like RNA, DNA and protein, which actively participate in the processes called transcription and translation. The central dogma of molecular biology refers to the collective process of transcription followed by translation, often called gene expression. The gene expression process is the foundation of all the chemical reactions in our body.

This process starts with DNA. As shown in Figure 1.7, double-stranded DNA replicates itself into two single-stranded DNA. This single-stranded DNA then converts into mRNA (messenger RNA). This process is called transcription. The alphabets or bases in mRNA are grouped together in a set of three. These sets are called codons. Each codon corresponds to a particular amino acid. tRNA (transfer RNA) translates these codons and transfers a free amino acid to the target. The target place where codon and amino acid bind takes place in the ribosome. Sequences of amino acids form polypeptide chains and thus proteins. Ribosomes then deliver the protein into the cell. These proteins are the building blocks of our body and are responsible for all the functions in our body.

1.4   THE FOLDING PROBLEM

The protein folding problem refers to the prediction of structures of proteins by analyzing the sequence of amino acids. In other words, if any of the twenty amino acids sequence together in a particular fashion, we should be able to judge how this polypeptide chain or amino acid sequence is going to fold and what will be the resultant proteins.

Further, polypeptide chains may experience interactions between two pairs that are distantly related but eventually came close due to massive folding of the

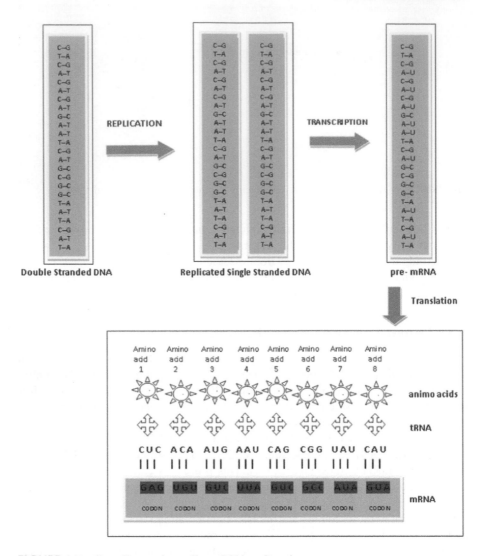

FIGURE 1.7 Gene Expression – From DNA to Protein

polypeptide sequence. These interactions are quite important in understanding the structure and function of proteins.

A lot of computer-based software are designed for the prediction of protein structures from sequences. These software take amino acid sequences as input and output protein structures.

With the sensation of automatic prediction of some of the proteins, designing our own protein has become possible. Things that were never seen in the biological world can now be made; for example, metacin, which can target the flu virus. This has brought new insights into the field of medicine.

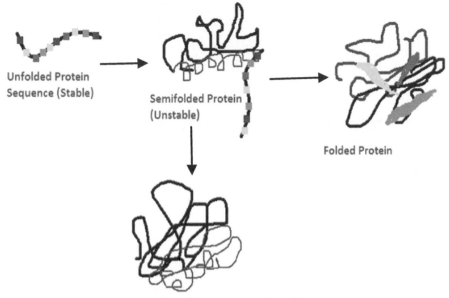

Unfolded Protein
Sequence (Stable)

Semifolded Protein
(Unstable)

Folded Protein

Toxic Protein Clump/ Incorrect Peptide Polymer

FIGURE 1.8   The Protein Folding Problem

The second issue related to protein folding is proteins that erroneously folded, as shown in Figure 1.8. If the folded protein structure deviates from the expected or actual structure, it leads to diseases in our body. The cause of formation of deviated structures may be environmental factors, or external stimuli in our body etc. Scientists and researchers have been working on this for decades. This can be controlled at the gene expression level. Genes may be allowed to under-express or over-express so as to control the structure of the resultant protein. Lots of therapeutic techniques have been developed for this. Consequently, this information is used to invent new drugs for targeted diseases.

1.5   GENOMICS AND PROTEOMICS

The journey from genotype to phenotype is characterized by genomics and proteomics. Genotype refers to all genetic material inside our body and phenotype refers to how we look on the outside of our body. As shown in Figures 1.9 and 1.10, RNA is synthesized from DNA, which in turn manufactures proteins. Protein–protein interaction leads to chemical reactions in our body that catalyze metabolism.

The aim of genomics and proteomics is the:

- Identification of proteins as targets to cure diseases.
- Understanding of cell processes and genetics.
- Study of evolution through phylogenetic trees.

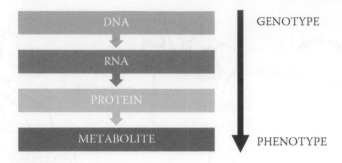

FIGURE 1.9   Genotype to Phenotype

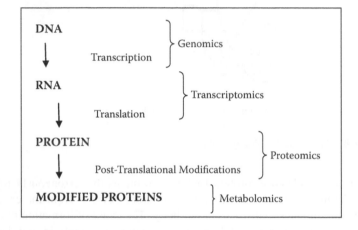

FIGURE 1.10   Genomics, Transcriptomics, Proteomics and Metabolomics

1.5.1   Genomics

The study of genes in the genome is called genomics. The entire genetic content or DNA of an organism is called genome. Each cell in our body contains base pairs that are formed from four bases of DNA (A, C, G, T). There are approximately 3.5 billion base pairs in a cell. The base-pair sequence folds multiple times to fit inside a cell. The most condensed form of this sequence is called a chromosome. We have twenty-three chromosomes in our body. Chromosomes contain genes, as shown in Figure 1.11. Massively coiled DNA exists in the form of chromosomes. Genes are the basic units of genetics. Each gene codes for a specific protein.

The aim of genomics is to:

- Study DNA sequences and its mutations.
- Study the flow of information within a cell.
- Diagnose new diseases and their possible treatments.
- Detect diseases in the body.
- Predict behavioural status based on gene expression.

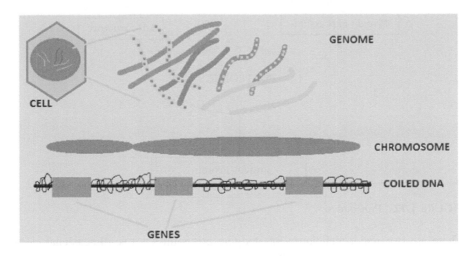

FIGURE 1.11   Gene, Chromosome, DNA and Genome

- Figure out how to resist diseases.
- Genetically modify food.
- Provide a comprehensive list of genes for the whole organism.
- Figure out information about the function of the DNA sequence, often called functional genomics.

1.5.2   Proteomics

The term *proteome* was coined in 1994 by Marc Williams. A collection of proteins in a cell is referred to as a proteome; the qualitative and quantitative study of proteomes of cellular organisms is proteomics.

Specifically, proteomics can answer the following:

- Identification of the structure of proteins
- Protein function
- Protein expression
- Protein post-translational modification
- Protein-protein interaction
- Protein localization

The protein-to-protein interaction is termed *proteomics;* in other words, the study of large sets of protein sequences is called proteomics.

There are more than 25,000 genes within the human genome. Each gene can produce more than one protein product. The total number of proteins in the human proteome is estimated to be more than one million.

The goal of proteomics is to study the dynamic protein product of genomes and their interactions. It leads to understanding cells, which are the units of life. Proteomics deals with the study of the interaction of RNA, DNA and proteins.

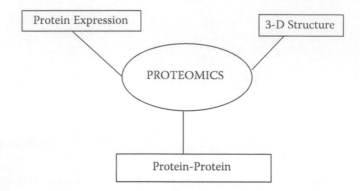

FIGURE 1.12   Proteomics

There are two major techniques for studying proteomics – two-dimensional electrophoresis and mass spectrometry.

Two-dimensional electrophoresis is the process of separation of complex protein
molecules. Mass spectrometry includes identification and structure analysis.

Proteomics can be classified as follows (Figure 1.12):

1. Protein Expression Comparison – Qualitative and quantitative study of
   protein expression between samples that differ by some variable.
2. Structural Proteomics – The goal of structural proteomics is to map the three-
   dimensional structure of protein and protein complexes.
3. Functional Proteomics – Functional proteomics aims to study protein-protein
   interactions, three-dimensional structure, cellular localization and post-
   translational modification in order to understand the functions of proteins.

APPENDIX

Exercise
Multiple-Choice Questions

1. The identification of drugs through genomic study is

   a. genomics.
   b. cheminformatics.
   c. pharmagenomics.
   d. phrmacogenetics.

2. The human genome contains about

   a. two billion base pairs.
   b. three billion base pairs.
   c. fout billion base pairs.
   d. five billion base pairs.

3. To synthesize a protein, a ribosome reads groups of three nucleotides called
   _____. Each of these represents one _____.

   a. amino acids; protein
   b. a reading frame; gene
   c. codons; amino acid
   d. DNA; allele

4. Below is the sequence of an extremely short mRNA:
   5' GCCAUUACAUGGUGCACUCCACCUAGGACUCACUC
   Show the DNA molecule from which this mRNA was transcribed.

   a. TACATGGTGCACTCCACCTAGGACTCACTC
   b. CGGTAATGTACCACGTGAGGTGGATCCTGAGTGAG

5. As the complexity of an organism increases, all of the following characteristics
   emerge except _____.

   a. the gene density decreases
   b. the number of introns increases
   c. the gene size increases
   d. an increase in the number of chromosomes
   e. repetitive sequences are present

6. The collection of proteins that can be produced by a given species is

   a. considered that species' genetic complement.
   b. correlated with the size of the organism.
   c. called the proteome.
   d. All of these.
   e. None of these.

7. The term *proteomics* was coined by

   a. Marc Wilkins.
   b. Marc Anthony.
   c. Marc Jacobs.
   d. Marc Wilkings.

8. 1 gene = 1 protein

   a. True
   b. False
   c. Can't say
   d. Rare situation

9. Proteins are made of amino acids linked together by specific bonds called

    a. peptide bonds.
    b. nitrogen bonds.
    c. hydrogen bonds.
    d. hydrogen and nitrogen bonds.

10. Which of the following is a correct definition of genetics?

    a. the study of transmission of traits from parent to offspring
    b. the study of genes and traits defined by genes
    c. the study of DNA
    d. the study of variation between members of a species

11. What is the main function of DNA?

    a. It stores information for protein synthesis.
    b. It can be mutated.
    c. It directs the process of protein synthesis.
    d. It provides energy for the cell.

12. Identify the correct order of the organization of genetic material, from largest to smallest.

    a. genome, chromosome, gene, nucleotide
    b. gene, chromosome, nucleotide, genome
    c. chromosome, gene, genome, nucleotide
    d. chromosome, genome, nucleotide, gene

13. Identify which of the following terms refers to the overall three-dimensional shape of a protein.

    a. primary structure
    b. secondary structure
    c. tertiary structure
    d. quaternary structure

14. Glucose is a

    a. protein.
    b. disaccharide.
    c. nucleic acid.
    d. monosaccharide.
    e. starch.

15. The process of copying a gene's DNA sequence into a sequence of RNA is called

    a. replication.
    b. transcription.
    c. translation.
    d. PCR.

16. Which molecule contains the genetic code?

    a. DNA
    b. mRNA
    c. tRNA
    d. rRNA

17. RNA contains which bases?

    a. adenine, thymine, guanine, cytosine, uracil
    b. adenine, thymine, guanine, cytosine
    c. thymine, guanine, cytosine, uracil
    d. adenine, guanine, cytosine, uracil

18. Which mode of information transfer usually does not occur?

    a. DNA to DNA
    b. DNA to RNA
    c. DNA to protein
    d. all occur in a working cell

19. Proteomics

    a. is another term for genomics in humans.
    b. is the study of the collection of proteins produced in a particular cell.
    c. is the study of proteins produced by a particular gene.
    d. proves that a single gene codes for only one protein.

20. A DNA strand with the sequence AACGTAACG is transcribed. What is the sequence of the mRNA molecule synthesized?

    a. AACGTAACG
    b. UUGCAUUGC
    c. AACGUAACG
    d. TTGCATTGC

Essay-Type Questions

Q1. Write short notes on the following:

    a. human genome
    b. DNA organization
    c. non-coding regions
    d. central dogma of molecular biology

Q2. a. How can we determine how active a gene is?

    b. Write short notes on the following:
       i.   Genomics and proteomics
      ii.   Central dogma of molecular biology

Q3. Bioinformatics is an interdisciplinary area. Support this statement.

Q4. a. "Proteomics bridges the gap between genomics and cellular information". Do you agree with this statement? Support your answer.

    b. How are genomics, transcriptomics, proteomics and metabolomics related?

Q5. What are nucleic acids? What are its types?

Q6. Differentiate between RNA and DNA.

Q7. What do you mean by gene expression? Can we control this process? If yes, support your answer.

Q8. How are gene, DNA and chromosome related?

Q9. Discuss various types of RNA and their functions.

Q10. Is protein formation always correct? If not, what is the impact of improper formation of proteins in our body?

2 Computational Biology

A few researchers use the terms *bioinformatics* and *computational biology* interchangeably. But these terms are actually different. Bioinformatics more often is related to a biology background. It deals with solving biological problems with the help of available tools. On the other hand, computational biologists tend to write algorithms and code for developing those tools that are to be used by bioinformaticists. In other words, computational biologists are more computer professionals than bioinformaticists.

In the previous chapter, we have learnt that there are three types of sequences:

- RNA sequence
- DNA sequence
- Protein sequence

In this chapter, we will learn about alignment of these sequences.

2.1 SEQUENCE ALIGNMENT

Sequence alignment is the process of placing two or more sequences one over the other such that regions of maximum similarity can be located, as shown below.

H	E	L	L	O	(Sequence 1)
H	E	L	P	_	(Sequence 2)
✓	✓	✓	✗	o	

Sequence 1 has five letters and sequence 2 has four letters. A checkmark (✓) signifies a match of letter. A cross (✗) signifies a mismatch and a circle (O) signifies a gap.

To summarize, we can say there can be three outcomes of similarity search.

1. Match
2. Mismatch
3. Gap

Another important aspect of sequence alignment is to discover regions of maximum similarity, as shown in the examples below.

U	C	A	U	G	(Sequence 1)
C	A	U	U	G	(Sequence 2)

Scenario 1

U	C	A	U	G		(Sequence 1)
C	A	U	U	G		(Sequence 2)
✗	✗	✗	✓	✓		

Scenario 2

U	C	A	U	G	_	(Sequence 1)
_	C	A	U	U	G	(Sequence 2)
o	✓	✓	✓	✗	o	

Scenario 3

U	C	A	_	U	G	(Sequence 1)
_	C	A	U	U	G	(Sequence 2)
o	✓	✓	o	✓	✓	

Scenario 4

U	C	A	U	_	G	(Sequence 1)
_	C	A	U	U	G	(Sequence 2)
o	✓	✓	✓	o	✓	

Scenario 1 shows two matches, three mismatches and zero gaps. Scenario 2 shows three matches, one mismatch and two gaps. Scenarios 3 and 4 show four matches, zero mismatches and two gaps. The score of sequence alignment is calculated on the basis of the predefined score function. For example, a score function may be defined as:

$$\text{Score} = \begin{cases} +2 & \text{if a match occurs} \\ -1 & \text{if a mismatch occurs} \\ -2 & \text{gap penalty} \end{cases}$$

Using the scoring function to score for scenario 1 is 1 (2+2-1-1-1). Similarly, the score of other scenarios may be calculated as 1, 4, and 4, respectively. So the best alignments are Scenarios 3 and 4.

2.2   GLOBAL SEQUENCE ALIGNMENT

Global sequence alignment refers to the procedure of aligning all the nucleotides of sequences. This method was proposed by Needleman, S. B. and Wunsch, C. D. in 1970. The algorithm is known as the Needleman-Wunsch algorithm.

A scoring matrix is created where the number of rows and columns correspond to the number of nucleotides (letter) in the two sequences, respectively. The rules for filling the matrix are as follows:

Step 1. Select a suitable score function. We are using this score function:

$$\text{Score} = \begin{cases} +2 & \text{if a match occurs} \\ -1 & \text{if a mismatch occurs} \\ -2 & \text{gap penalty} \end{cases}$$

	C2					
C	-6					
A	-5					
T	-4					
C	-3					
A	-2					
G	-1					
∽(gap)	0	-1	-2	-3	-4	R7
		∽	A	C	G	C

FIGURE 2.1 Global Sequence Alignment-Step 1

Step 2. Rules for filling the rest of the matrix:

- Adjacent box (add gap)
- Bottom box (add gap)
- Diagonal box (consider match and mismatch)

Step 3. Insert a gap in the first column and last row, as shown in Figure 2.1.

Step 4. At the intersection of the two gaps, put a zero.

Step 5. The second column (C2) and second-to-last row (R7) are filled by consecutively adding a gap value to adjacent cells, as shown in Figure 2.1.

Step 6. Consider a small square of matrix, shown in Figure 2.2. This small square is as shown in Figure 2.3.

The value in the highlighted cell is filled by considering the values that are coming from the horizontal and vertical direction as well as from diagonal direction.

Step 7. As mentioned, the value of the gap is added while moving in the horizontal or vertical direction. Therefore, the resultant value coming from the horizontal direction is calculated as:

Resultant Value = value from beside box + gap value

$$= -1 + (-1)$$
$$= -2$$

Step 8. Similarly, the resultant value coming from the vertical direction is calculated as:

C	-6				
A	-5				
T	-4				
C	-3				
A	-2				
G	-1				
∽	0	-1	-2	-3	4
	∽	A	C	G	C

FIGURE 2.2 Global Sequence Alignment-Step 2

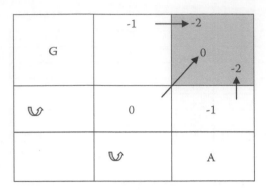

FIGURE 2.3   Global Sequence Alignment-Step 3

Resultant Value = value from bottom box + gap value
$$= -1 + (-1)$$
$$= -2$$

Step 9. Finally, the resultant value coming diagonally is calculated as:

Resultant Value = value from diagonal box + match/mismatch value
$$= 0 + 0$$
$$= 0$$

As G and A letters are mismatched, the score of the mismatch i.e. zero will be added to the value from the diagonal box.

Step 9. Now consider the highest out of these three resultant values i.e. 2, 0, and 2. The maximum value is zero so zero is saved as the final value for the highlighted cell. Also, the direction, from which the final value came, is also saved for the trace-back step. The resultant matrix is shown in Figure 2.4.

Step 10. Steps 5 to 9 are repeated to fill the matrix completely, as shown in Figure 2.5.

C	-6				
A	-5				
T	-4				
C	-3				
A	-2				
G	-1	0			
↰	0	-1	-2	-3	-4
	↰	A	C	G	C

FIGURE 2.4   Global Sequence Alignment-Step 4

C	▲-6	▲ -4	↗ -2	↗ -1	↗ 1
A	▲-5	-3	▲ -1	↗ 0	↗ 1
T	▲-4	▲ -2	↓ 0	↗ 1	↗ 0
C	▲-3	-1	↓ 1	↗ 0	↗ 0
A	▲-2	↗ 0	↗ 0	↗ -1	↗ -1
G	▲-1	↗ 0	↗ -1	↗ -1	→ -2
↻	0	↙ -1	↙ -2	→ -3	→ -4
	↻	A	C	G	C

FIGURE 2.5 Global Sequence Alignment-Step 5

Step 11. The next step is to trace back. Find out the highest value in the matrix. In our example, the highest value is 1. Generally, the highest value lies in the last column.

Step 12. Now trace back to the box from where this value came from. In our example, the value of cell (C6, R1) came from cell (C5, R2), as shown in Figure 2.5. The further value from cell (C5, R2) came from cell (C4, R3). Similarly, trace back to cell (C2, R6). The trace-back is shown in Figure 2.6.

Step 13. After tracing back, the next step is alignment of sequences. The rules for alignment are as follows:

- For diagonal arrows in tracing back, put letters during alignment.
- For horizontal arrows, put a gap during alignment.

Step 14. The arrow from cell (C6, R1) to cell (C5, R2) is diagonal, so the corresponding letters of cells (C1, R1) and (C6, R8), respectively, are aligned and lead to matching letters.

Step 15. The next arrow from (C5, R2) to cell (C4, R3) is again a diagonal, so corresponding letters of cells (C1, R2) and (C5, R8), respectively, are aligned and lead to a mismatch.

Step 16. On the contrary, the next arrow from (C4, R3) to cell (C4, R4) is a vertical arrow, so the gap is written as shown below:

C1	C2	C3	C4	C5	C6	
C	-6	-4	-2	-1	↗ 1	R1
A	-5	-3	-1	↗ 0 ◤	1	R2
T	-4	-2	↓ 0 ◤	1	0	R3
C	-3	-1	↘ 1	0	0	R4
A	-2	↗ 0 ◤	0	-1	-1	R5
G	-1	◤ 0	-1	-1	-2	R6
↻	0	-1	-2	-3	-4	R7
	↻	A	C	G	C	R8

FIGURE 2.6 Global Sequence Alignment-Step 6

G	A	C	T	A	C
–	A	C	–	G	C

Step 17. The alignment is completed.

2.3   LOCAL SEQUENCE ALIGNMENT

Contrary to global sequence alignment, local sequence alignment refers to the alignment of a segment of the sequences and not the complete sequences.

Similarity in sequences is often correlated to homology. Homologous sequences are the ones that belong to the same ancestor. Thus, structure or function can be inferred from sequence alignments. It may not be necessary to align the complete sequences; rather, we may need to align only parts of sequences. In that case, local sequence alignment is a better choice.

The local alignment method was proposed by Smith, T. F. and Waterman, M. S. in 1981. The algorithm is popularly known as the Smith-Waterman algorithm.

The difference between the Needleman-Wunsch algorithm and Smith-Waterman algorithm lies in the filling matrix. The scoring function as well as rules for filling the matrix remain the same. The only difference lies in the values of a matrix. Negative numbers are written as zero and positive numbers are written as is. Another slight difference is in the trace-back step. Here, we will start with the highest number and trace it diagonally until we reach a cell with a zero value, as shown in Figure 2.7.

The resultant alignment is as follows:

A	T	C
C	G	C

Local sequence alignment is important when:

- We need to align two partially overlapping sequences
- One sequence is a subset of another sequence

C	0	0	0	0	1
A	0	0	0	0	1
T	0	0	0	1	0
C	0	0	1	0	0
A	0	0	0	0	0
G	0	0	0	0	0
ᴗ	0	0	0	0	0
	ᴗ	A	C	G	C

FIGURE 2.7   Local Sequence Alignment

- The sequence length of two sequences to be aligned varies too much
- The goal is to find conserved regions among sequences

2.4   MULTIPLE SEQUENCE ALIGNMENT

Local sequence alignment as well as global sequence alignment is collectively known as pairwise alignment. Sometimes we require alignment of more than two sequences; for instance, the number of sequences may be 10, 20 or maybe 60. As the number of sequences to be aligned increases, the complexity of the alignment algorithm increases.

An application of multiple sequence alignment may be construction of a phylogenetic tree. A phylogenetic tree is the one that shows how closely or distantly organisms are related. For constructing such a tree, conserved regions need to be found. Conserved regions may be located with the help of the multiple sequence alignment. Pairwise alignment may not guarantee an ancestral relationship between sequences of two species, but ancestral relationships may be significantly analyzed with the multiple sequence alignment.

Pairwise sequence alignment requires a two-dimensional matrix. Similarly, sequence alignment of three sequences will require a three-dimensional matrix. Further, a four-dimensional sequence will be used to align four sequences, and so on.

For example, we take three sequences – AGCT, GGT and CGCT – as shown in Figure 2.8.

The vertices to be plotted on a three-dimensioanl matrix are (1,0,0), (1,0,1), (2,1,2), (2,2,2), (3,2,3) and (4,3,4).

Further, the values in cells of a two-dimensional matrix used to come from three directions, diagonally, horizontally and vertically. But, in the case of an n-dimensional matrix, values will come from various directions into cells of the score matrix, as shown in Figure 2.9.

A multiple sequence alignment can be achieved using alignment tools such as ClustalX, ShadyBox, PrettyBox etc. Following are the steps to conduct a multiple sequence alignment:

I	1	1	2	2	3	4
	A	-	G	-	C	T

J	0	0	1	2	2	3
	-	-	G	G	-	T

k	0	1	2	2	3	4
	-	C	G	-	C	T

FIGURE 2.8   Multiple Sequence Alignment

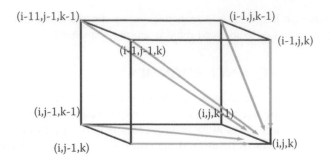

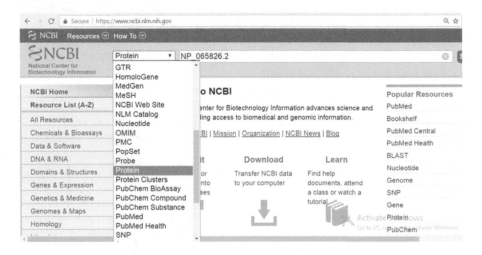

FIGURE 2.9   n-Dimensional Matrix for Multiple Sequence Alignment

FIGURE 2.10   NCBI – Multiple Sequence Alignment

Step 1. Open NCBI (https://www.ncbi.nlm.nih.gov/) and obtain protein sequences, as shown in Figure 2.10.

Step 2. After obtaining the protein sequences, run Blast to find similar sequences, as shown in Figures 2.11–2.14.

Step 3. Select taxonomy reports to generate organism reports. Randomly select the accession numbers of generated sequences, as shown in Figure 2.15.

Step 4. Now search the selected sequences on NCBI (Figure 2.16) home page, as used in step 1.

Step 5. The resultant sequences are shown in Figure 2.17.

Step 6. Sequences can be saved in FASTA format or FASTA (text) format by clicking "Summary", as shown in Figures 2.18 and 2.19.

Step 7. Now copy the FASTA format of all the sequences in a multiple alignment tool, such as Clustal Omega (http://www.ebi.ac.uk/Tools/msa/clustalo/) and run multiple sequence alignments. The result is shown in Figures 2.20 and 2.21.

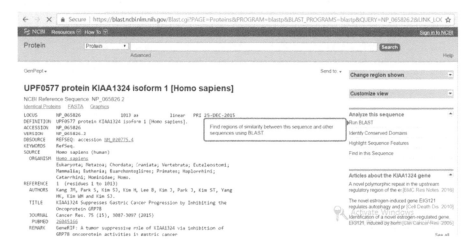

FIGURE 2.11   NCBI – Multiple Sequence Alignment

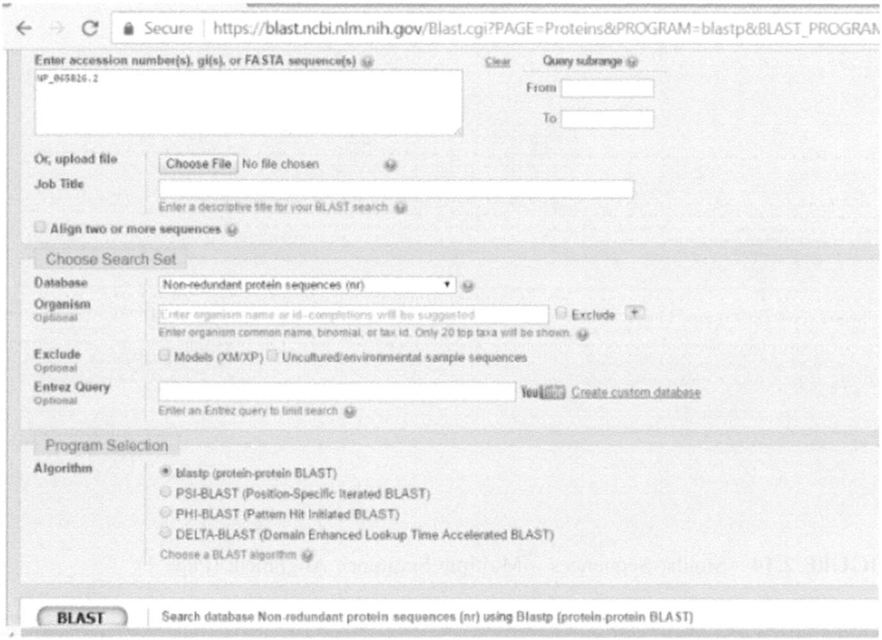

FIGURE 2.12   BLAST – Multiple Sequence Alignment (Step 1)

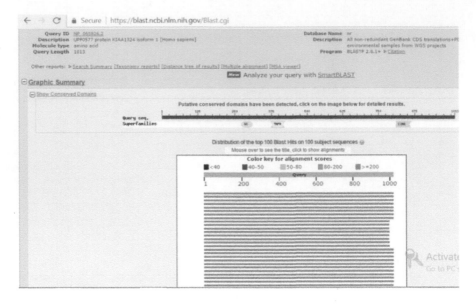

FIGURE 2.13   BLAST – Multiple Sequence Alignment (Step 2)

FIGURE 2.14   Similar Sequences – Multiple Sequence Alignment (Page 1)

FIGURE 2.15   Similar Sequence – Multiple Sequence Alignment (Page 2)

FIGURE 2.16   NCBI – Multiple Sequence Alignment (Step 1)

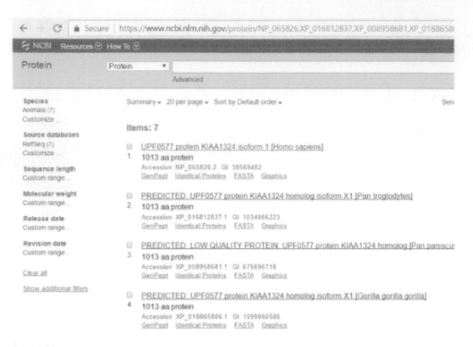

FIGURE 2.17   NCBI – Multiple Sequence Alignment (Step 2)

FIGURE 2.18   NCBI – Multiple Sequence Alignment (Step 3)

FIGURE 2.19 NCBI – Multiple Sequence Alignment (Step 4)

FIGURE 2.20 Multiple Sequence Alignment

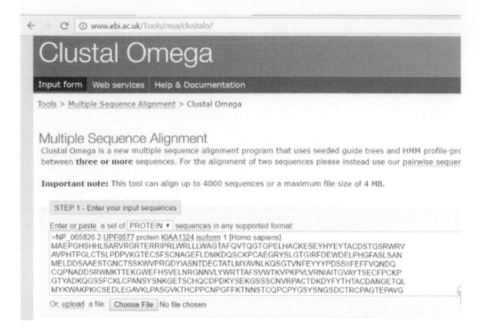

FIGURE 2.21   Clustal Omega – Multiple Sequence Alignment Example

FIGURE 2.22   Clustal Omega – Multiple Sequence Alignment

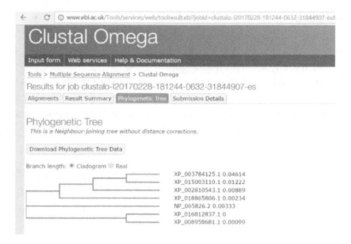

FIGURE 2.23   Cladogram – Clustal Omega Phylogenetic Tree

Step 8. The cladogram or phylogenetic tree can now be constructed by clicking "Phylogenetic Tree" as shown in Figures 2.22 and 2.23.

APPENDIX

Implementing the Needleman-Wunsch global sequence alignment algorithm using ASP.net and C#

Sequence alignment is a way of arranging two or more sequences of characters to identify regions of similarity – similarities may be a consequence of functional or evolutionary relationships between these sequences.

It is a procedure for comparing two or more sequences by searching for a series of individual characters that are in the same order in those sequences.

Find optimal global alignment between sequences for a given set of rules.

There are three steps:

1. Initialization step
2. Matrix fill step
3. Trace-back step

Design front end

1. Take five input boxes and one button, first two text boxes for sequences, three text boxes for match, mismatch and gap value and button to perform a click action.
2. Take two gridviews and labels to visualize the matrix operation and output both sequence alignments.

C# code explained

1. Input five variables for the length of the first sequence, length of the second sequence, match, mismatch and gap.
2. Define two matrices of first and second sequence length + 1, one for scoring and the second for trace-back.
3. Fill both matrices according to the score value (max score value is opt).
4. In the traceback matrix, use D for diagonal, L for left, U for upper score value.
5. Convert both matrices into a data table to visualize the data into a grid view.
6. Call the trace-back function to display both aligned sequences in labels.
7. If the value is D, i.e. diagonal, then write both characters of both sequences; otherwise, in the case of L and U score values, we write only one character of the corresponding sequence.

Code of ASP.CS page

```
using System;
using System.Collections.Generic;
using System.Data;
using System.Web;
using System.Web.UI;
using System.Web.UI.WebControls;
public partial class GlobalSeq: System.Web.UI.Page
{
    protected void btnFillMatrix_Click(object sender, EventArgs e)
    {
    int a = txtSeq1.Text.Length+1;
    int b = txtSeq2.Text.Length+1;
    int ma=Convert.ToInt32(txtMatch.Text);
    int mi=Convert.ToInt32(txtMisMatch.Text);
    int Ga=Convert.ToInt32(txtGap.Text);
    int[,] ScoreMatrix=new int[a,b];
    char[,] tracebackMatrix = new char[a, b];
    ScoreMatrix[0,0]=0;
    for (int i = 1; i < a; i++)
    {
        ScoreMatrix[i, 0] = i * Ga;
        tracebackMatrix[i, 0] = 'U';
    }
    for (int j = 1; j < b; j++)
    {
        ScoreMatrix[0, j] = j * Ga;
        tracebackMatrix[0, j] = 'L';
    }
    for (int i = 1; i < a; i++)
```

```
{
    for (int j = 1; j < b; j++)
    {
    int scroeDiag = 0;
    if (txtSeq2.Text.Substring(j - 1, 1) == txtSeq1.Text.Substring(i - 1, 1))
        scroeDiag = ScoreMatrix[i - 1, j - 1] + ma;
    else
        scroeDiag = ScoreMatrix[i - 1, j - 1] + mi;
        int scroeLeft = ScoreMatrix[i, j - 1] +Ga;
        int scroeUp = ScoreMatrix[i - 1, j] +Ga;
        int maxScore = Math.Max(Math.Max(scroeDiag, scroeLeft), scroeUp);
        ScoreMatrix[i, j] = maxScore;
        if (ScoreMatrix[i, j] == scroeDiag)
        {
        tracebackMatrix[i, j] = 'D';
        }
        else if (ScoreMatrix[i, j] == scroeLeft)
        {
        tracebackMatrix[i, j] = 'L';
        }
        else if (ScoreMatrix[i, j] == scroeUp)
        {
        tracebackMatrix[i, j] = 'U';
    }
    }
}
GridView1.DataSource = ConvertArrayToDataTable(ScoreMatrix, a, b);
GridView1.DataBind();
GridView2.DataSource = ConvertArrayToDataTable(tracebackMatrix, a, b);
GridView2.DataBind();
TraceBack(tracebackMatrix, txtSeq1.Text, txtSeq2.Text);
}
public DataTable ConvertArrayToDataTable(int[,] ScoreMatrix,int a,int b)
{
DataTable dt = new DataTable();
for (int j = 0; j < b; j++)
    dt.Columns.Add(j.ToString());
for (int i = 0; i < a; i++)
{
    dt.Rows.Add();
    for (int j = 0; j < b; j++)
    dt.Rows[i][j] = ScoreMatrix[i, j];
}
dt.AcceptChanges();
return dt;
```

```csharp
    }
    public DataTable ConvertArrayToDataTable(char[,] ScoreMatrix, int a, int b)
    {
    DataTable dt = new DataTable();
    for (int j = 0; j < b; j++)
        dt.Columns.Add(j.ToString());
    for (int i = 0; i < a; i++)
    {
        dt.Rows.Add();
        for (int j = 0; j < b; j++)
        dt.Rows[i][j] = ScoreMatrix[i, j];
    }
    dt.AcceptChanges();
    return dt;
    }
    public void TraceBack(char[,] tracebackMatrix, string sequenzA, string
sequenzB)
    {
        int i = tracebackMatrix.GetLength(0) - 1;
        int j = tracebackMatrix.GetLength(1) - 1;
        string alignedSeqA = "";
        string alignedSeqB = "";
        while (i!= 0 || j!= 0)
        {
        switch (tracebackMatrix[i, j])
        {
            case 'D':
            alignedSeqA += sequenzA[i - 1];
            alignedSeqB += sequenzB[j - 1];
            i--;
            j--;
            break;
        case 'U':
            alignedSeqA += sequenzA[i - 1];
            alignedSeqB += "-";
            i--;
            break;
        case 'L':
            alignedSeqA += "-";
            alignedSeqB += sequenzB[j - 1];
            j--;
            break;
        }
        }
        lblSequence1.Text = reverseString(alignedSeqA);
```

```
        lblSequence2.Text = reverseString(alignedSeqB);;
    }
    public string reverseString(string strInp)
    {
        char[] arr = strInp.ToCharArray();
        Array.Reverse(arr);
        return new string(arr);
    }
}
```

Output screenshot

Output for two given sequences, ACGC and GACTAC

APPENDIX

Exercise
  Multiple-Choice Questions

Q1. Which programming technique is used in the Needleman-Wunsch algorithm
     to allow it to efficiently carry out global alignment?

     a. heuristic approach
     b. list comprehension
     c. dynamic programming
     d. matrix algebra

Q2. If two sequences show significant similarity, it means

     a. they are paralogs.
     b. they are orthologs.
     c. they evolved from a common ancestor.
     d. HGT has occurred.

Q3. Which of the following is a sequence alignment tool?

     a. BLAST
     b. PRINT
     c. PROSITE
     d. PIR

Q4. A multiple sequence alignment of related genes can identify amino acids
     required for protein function.

     a. True
     b. False

Q5. The BLAST program is used in

     a. DNA sequencing.
     b. amino acid sequencing.
     c. DNA barcoding.
     d. bioinformatics.

Q6. Clustal W is a

     a. multiple sequence alignment tool.
     b. protein secondary structure predicting tool.
     c. data retrieving tool.
     d. nucleic acid sequence analysis tool.

Q7. The procedure of aligning two sequences by searching for patterns in the same order in the sequences is

    a. sequence alignment.
    b. pairwise alignment.
    c. multiple sequence alignment.
    d. all of these.

Q8. Sequence alignment helps scientists

    a. trace out evolutionary relationships.
    b. infer the functions of newly synthesized genes.
    c. predict new members of gene families.
    d. with all of these.

Q9. Well-conserved regions in multiple sequence alignments

    a. reflect areas of structural importance.
    b. reflect areas of functional importance.
    c. reflect areas of both functional and structural importance.
    d. reflect areas likely to be of functional and/or structural importance.

Q10. Why are colour schemes important in creating and analyzing sequence alignments?

    a. They look pretty.
    b. They make clearer printouts and presentations.
    c. They allow you to distinguish conserved residues and residue groups more easily.
    d. They allow you to detect active sites of proteins.

Essay-Type Questions

Q1. What do you mean by sequence alignment? What are its applications in computational biology?

Q2. Explain global sequence alignment in detail. Also, perform global sequence alignment on the below-mentioned DNA sequences:

T G A
T T A

Q3. Differentiate between the procedure of carrying out global sequence alignment and local sequence alignment.

Q4. Align the below-mentioned sequences:

ACTGATTCA
ACGCATCA

Q5. When is local sequence alignment important?

3 Phylogenetics

3.1 INTRODUCTION TO PHYLOGENETICS

The term *species* is defined as a set of similar organisms. The degree of similarity is such that the amount of dissimilarity to other sets is significant. The similarity accounts for gene flow among the species. In other words, species grow by producing viable offsprings. Organisms that are isolated from this genetic transfer experience a drift from their species with due course of time. This is called speciation. Formally, speciation may be defined as an evolutionary process that leads to formation of new species. For example, a group of lizards may happen to move outside their native land. With due course of time, those lizards try to adapt to the environment in order to sustain life in the new environment. These lizards may undergo structural changes like change in colorful flap of skin under the lizard's throat, leading to new species of lizards that no longer mate with the former species of lizards. Over a period of hundreds of years, the old species may not even recognize the new species, thereby creating biological taxonomy.

This biological taxonomy relates to term *phylogeny*. Consequently, phylogenetics is a science of studying biological relationships between organisms that have emerged from a common ancestor over a period of time.

Phylogenetic relationships are represented by a tree diagram or a hierarchical diagram called a phylogenetic tree. In other words, a phylogenetic tree is a method of showing the genetic relatedness between species. It portrays the evolutionary relationships that are developed over a period of time from a common ancestor. Figure 3.1 shows a phylogenetic tree wherein Taxa X and Taxa Y have a common ancestor. X, Y and Z are called leaves of the tree. Taxa X and Taxa Y are more closely related to each other than to Taxa Z because they share a common ancestor. But this doesn't mean that Taxa X, Y and Z are not related. They are very much related because they also share a common ancestor at the second level. This way we can say a phylogenetic tree shows how closely or distantly the organisms are related.

There are two types of phylogenetic trees: rooted phylogenetic trees and unrooted phylogenetic trees.

3.1.1 ROOTED PHYLOGENETIC TREE

A rooted phylogenetic tree is as shown in Figure 3.1. Organisms here are clearly related to each other by a common ancestor. Evolutionary relationships can be built by means of physiological features, morphological features and significantly the genetic features of each of the organisms. Physiological features relate to the functioning of a particular organ; for example, how an organism's metabolism works.

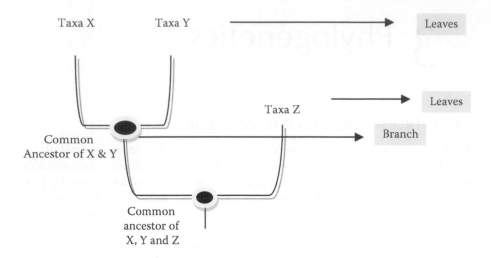

FIGURE 3.1   Rooted Phylogenetic Tree

Morphological features relate to various forms (morph) of an organ and other structural features for example structures of bones. Genetic features of an organism relate to features that are generally transmitted from parent to child; for example, colour of the allele of the eye of a mother and child.

3.1.2   UNROOTED PHYLOGENETIC TREE

Unlike rooted phylogenetic trees, unrooted trees are omnidirectional. More precisely, it is difficult to find ancestor of organisms by merely looking at the unrooted tree. In other words, an unrooted tree is constructed when we are not confident about the ancestor of two or more organisms. An unrooted tree is shown in Figure 3.2 in which X, P, Y, Z. An unrooted try can also be transformed into a rooted try if additional information related to morphology, physiology or genetics is available. The resultant rooted tree may not always be correct due to lack of additional information.

3.2   MOLECULAR PHYLOGENETICS

Molecular phylogenetics is a branch of phylogeny that analyzes the difference in molecular sequences, essentially in a DNA sequence to understand the evolutionary relationships. Over a period of time, molecular sequences observe mutations.

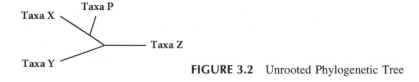

FIGURE 3.2   Unrooted Phylogenetic Tree

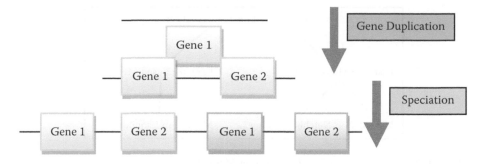

FIGURE 3.3   Paralogs and Orthologs

In molecular phylogenetics, the rate of mutation in a molecular sequence is tracked in order to study evolutionary relationships among species. A mutation in a sequence refers to insertion, removal, inversion or replacement of nucleotides in a molecular sequence.

Information is passed from generation to generation through gene evolution or through DNA sequence. Gene evolution results in paralogs, homologs and orthologs. Paralogs are homologous sequences that are related due to gene duplication, as shown in Figure 3.3. Orthologs are homologous sequences that are generated via speciation. In other words, orthologs have a last common ancestor. Homologs may result from speciation or gene duplication. Homologs may be paralogs or orthologs.

The molecular structure of related species has a high degree of similarity.

3.3   PHYLOGENETIC TREES AND THEIR CONSTRUCTION

Understanding the phylogenetic relationships between organisms aids in creation of a phylogenetic tree, as shown in Figure 3.4.

Phylogenetic tree construction methods may be categorized, as shown in Figure 3.5.

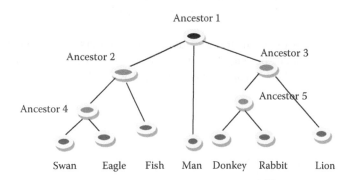

FIGURE 3.4   Phylogenetic Tree

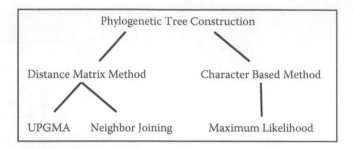

FIGURE 3.5   Phylogenetic Tree Construction Methods

3.4   DISTANCE MATRIX METHODS OF PHYLOGENETIC TREE CONSTRUCTION

Distance-based matrix methods use matrix as a data structure to conduct phylogenetic tree construction. The data is transformed into pairwise distances.

3.4.1   UNWEIGHTED PAIR GROUP METHOD WITH ARITHMETIC MEAN (UPGMA) – DISTANCE MATRIX METHOD

UPGMA is based on an agglomerative clustering technique. In this method, closely related clusters are progressively combined to form one cluster. Combining clusters carries on until all the clusters are combined to form a tree structure. Let us take an example of four nodes, namely a, b, c and d, as shown in Figure 3.6.

The distance between the nodes is reflected in Table 3.1 as a two-dimensional matrix.

Step 1. Find two nodes that are closest to each other. In the above example, the closest nodes are c and b, with a distance of three units.

Step 2. Combine nodes b and c, as shown in Table 3.2, to form cluster bc.

$$\text{Distance (a to bc)} = (7 + 5)/2 = 6$$
$$\text{Distance (d to bc)} = (8 + 8)/2 = 8$$
$$\text{Distance (e to bc)} = (5 + 15)/2 = 10$$

Step 3. Create a phylogenetic tree according to new clusters, as shown in Figure 3.7. The new cluster is node "bc" and the distances from node "bc" to node "b" and from node "bc" to node "a" are equal. So, the distance 3 is divided into 1.5 each.

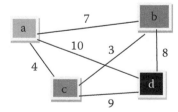

FIGURE 3.6   UPGMA Method Example

TABLE 3.1

UPGMA Example: Distance between Nodes/Clusters

	a	b	c	d	e
a	–	–	–	–	–
b	7	–	–	–	–
c	5	3	–	–	–
d	10	8	8	–	–
e	8	5	15	7	–

TABLE 3.2

UPGMA Example: Updated Distance between Nodes/ Clusters (Step 2)

	a	Bc	d	E
a	–	–	–	
bc	6	–	–	
d	10	8	–	
e	8	10		–

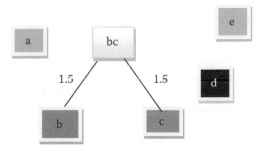

FIGURE 3.7 UPGMA Method Example: Step 3

Step 4. Repeat steps 1 to 3 until single distance value remains in matrix. Now consider Table 3.2. The minimum distance value is six. Therefore nodes "a" and "bc" will be clustered as shown in Table 3.3 and Figure 3.8.

$$\text{Distance (d to abc)} = (10 + 8)/2 = 9$$
$$\text{Distance (a to abc)} = (10 + 8)/2 = 9$$

Step 5. Similarly, the calculations may be as follows.

$$\text{Distance (de to abc)} = (9 + 9)/2 = 9$$

TABLE 3.3

UPGMA Example: Updated Distance between Nodes/Clusters (Step 4)

	abc	d	E
abc	0		
d	9	0	
e	9	7	0

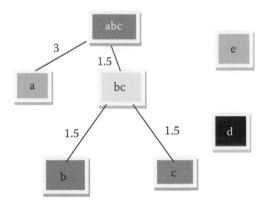

FIGURE 3.8 UPGMA Method Example: Step 4

Let us consider the tree once again. The distance of cluster "abc" is six which is divided into two halves amongst "a" and "bc". Now the total distance from node "abc" to each leaf node should be 3. Distance from node "bc" to node "b" is already 1.5. So the remaining distance will be,

Distance (abc to bc) = distance (abc to b) – distance (bc to b) = 3 – 1.5 = 1.5

Step 6. From Table 3.3, it is clear that nodes "d" and "e" will b clustered. The resultant tree and corresponding table are shown in Figure 3.9 and Table 3.4.

Step 7. By repeating the procedure, the phylogenetic tree is as generated in Figure 3.10.

It may be noted that the sum of distances from each parent node to their respective leaves is the same. The distance from the root node to all leaves is same.

This problem has been observed as the major disadvantage of the UPGMA algorithm as the number of mutations from one taxa to another may be different in the real world.

To incorporate variable mutations, a new method was proposed – neighbor joining method.

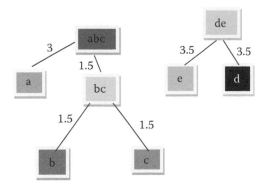

FIGURE 3.9   UPGMA Method Example: Step 6

TABLE 3.4
UPGMA Example: Updated Distance between Nodes/Clusters (Step 6)

	abc	De
abc	0	
de	9	0

3.4.2   NEIGHBOR JOINING METHOD – DISTANCE MATRIX METHOD

The neighbor joining method is also a distance-based clustering method, but it generates a phylogenetic tree with a variable length from the parent node to the leaves. It generates an unrooted tree.

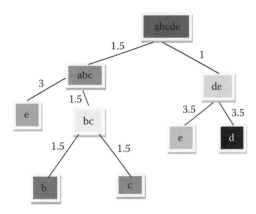

FIGURE 3.10   UPGMA Method Example: Step 7

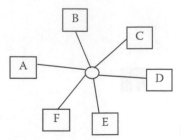

FIGURE 3.11   Neighbor Joining Method Example

Step 1. For input, create a random tree using a star topology, as shown in Figure 3.11.

Step 2. Find out the closest nodes and join them.

Step 2.1. For each of the nodes, calculate u_i such that $u_i = \Sigma_j\ D_{ij}\ /\ (n-2)$ where D_{ij} (Table 3.5) is the distance between nodes i and j, as given in the matrix (Figure 3.12) and n is the total number of nodes. The resultant matrix is:

Step 2.2. Calculate the distance matrix using this formula:

$$D_{ij} - u_i - u_j$$

For example, for nodes A and B, the value is calculated as:

TABLE 3.5

Neighbor Joining Method Matrix

I	$u_i = \Sigma_j\ D_{ij}/(n-2)$
A	$(16 + 20 + 26)/(4 - 2) = 31$
B	$(16 + 11 + 17)/(4 - 2) = 22$
C	$(20 + 11 + 13)/(4 - 2) = 22$
D	$(26 + 17 + 13)/(4 - 2) = 28$

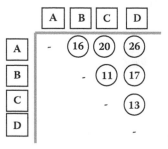

FIGURE 3.12   Neighbor Joining Method Example (Step 2.1)

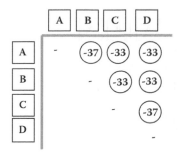

FIGURE 3.13 Neighbor Joining Method Example (Step 2.2)

$$D_{AB} - u_A - u_B = 16 - 31 - 22 = -37$$

The resultant distance matrix is shown in Figure 3.13.

Select the nodes that are closest to each other. In this case, A, B and C, D are equidistant. So, we randomly choose C, D as the closest neighbors.

Step 3. Calculate V_i and V_j as:

$$V_i = 0.5 \ D_{ij} + 0.5 \ (u_i - u_j)$$
$$V_j = 0.5 \ D_{ij} + 0.5 \ (u_j - u_i)$$

For example:

$$V_C = 0.5 \ D_{CD} + 0.5 \ (u_C - u_D) = 0.5 * 13 + 0.5 * (22 - 28) = 3.5$$
$$V_D = 0.5 \ D_{CD} + 0.5 \ (u_D - u_C) = 0.5 * 13 + 0.5 * (28 - 22) = 9.5$$

Step 4. The distance between nodes C and D is now divided into two parts – 3.5 and 9.5 –respectively. The phylogenetic tree construction may start now. The generated tree is shown in Figure 3.14.

Step 5. Next step is to combine nodes C and D in distance matrix. Also distances of other nodes from X need to be calculated as shown in Figure 3.15(a) and 3.15(b).

$$D_{XA} = (D_{CA} + D_{DA} - D_{CD})/2$$
$$= (20 + 26 - 13)/2 = 16.5$$

$$D_{XB} = (D_{CB} + D_{DB} - D_{CD})/2$$
$$= (11 + 18 - 13)/2 = 8$$

FIGURE 3.14 Neighbor Joining Method Example: Generated Tree (Step 4)

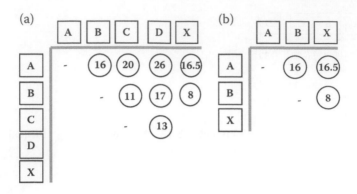

FIGURE 3.15   Neighbor Joining Method Example (Step 5)

Step 6. Repeat steps 2 to 4 until only one node is left in the matrix (Figures 3.16, 3.17 and 3.18) (Table 3.6).

$$V_A = 0.5 * 16 + 0.5 * (32.5 - 24)$$
$$= 12.25$$

$$V_B = 0.5 * 16 + 0.5 * (24 - 32.5)$$
$$= 3.75$$

$$D_{XY} = (D_{AX} + D_{BX} - D_{AB})/2$$
$$= (16.5 + 8 - 16)/2 = 4.25$$

The intermediate versions of phylogenetic trees are shown in Figure 3.17 and the final tree is shown in Figure 3.19.

The distances between nodes in the resultant phylogenetic tree can be cross verified from the initial distance matrix. For instance, $D_{AB} = 12.25 + 3.75 = 16$. The

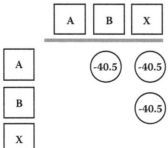

FIGURE 3.16   Neighbor Joining Method Example (Step 6)

FIGURE 3.17   Neighbor Joining Method Example: Generated Tree (Step 6)

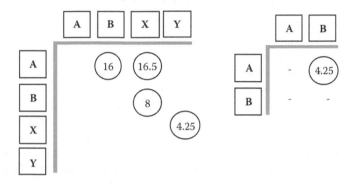

FIGURE 3.18   Neighbor Joining Method Example (Step 5)

TABLE 3.6
Neighbor Joining Method Matrix (Step 6)

I	$u_i - \Sigma_j\ D_{ij}/(n-2)$
A	$(16 + 16.5)/(3 - 2) = 32.5$
B	$(16 + 8)/(3 - 2) = 24$
X	$(16.5 + 8)/(3 - 2) = 24.5$

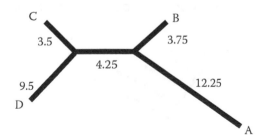

FIGURE 3.19   Neighbor Joining Method Example: Generated Final Tree

distance between nodes A and B is also 16 in the initial matrix. In similar fashion, other distances can be cross verified.

The advantages of this method are its fast processing and polynomial time algorithm. But this method generates an unrooted tree, so the common ancestor cannot be predicted.

3.5   MAXIMUM LIKELIHOOD – CHARACTER-BASED METHOD FOR PHYLOGENETIC TREE CONSTRUCTION

Character-based methods use the aligned characters, such as DNA or protein sequences, for constructing phylogenetic trees.

For instance, consider aligned DNA sequences of species A to E (Figure 3.20(a)). The transition (or evolutionary steps) in the first set of characters may result in a phylogenetic tree, as shown in Figure 3.20(b).

The input of this method is an n * m character matrix, where n denoted species and m depicts characters. The output is a tree in which species with similar character values appear near to each other. Characteristics could be number of fingers, number of walking legs, presence or absence of hair, etc. The characteristic data is placed in a character state matrix; the matrix is shown in Figure 3.21. The first step in a phylogenetic tree generation is arrangement of the matrix contents in such a way that it conveys some relationship between characteristics of species. Also, the number of evolutionary steps should be kept to a minimum while creating the tree.

Figure 3.21 shows a character state matrix for various taxa (lancelet, lamprey, tuna, salamander, turtle, leopard) and their corresponding characteristics. namely hair, amniotic egg, four walking legs, hinged jaws and vertebral column. The taxa leopard has all five characteristics and the taxa lancelet has none of the given characteristics. So, these two taxa will appear at the extreme ends in the

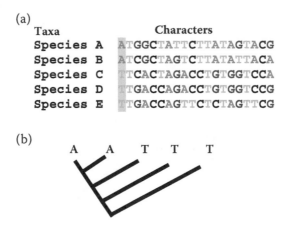

(a)

Taxa	Characters
Species A	ATGGCTATTCTTATAGTACG
Species B	ATCGCTAGTCTTATATTACA
Species C	TTCACTAGACCTGTGGTCCA
Species D	TTGACCAGACCTGTGGTCCG
Species E	TTGACCAGTTCTCTAGTTCG

(b)

A   A   T   T   T

FIGURE 3.20   (a) DNA Sequences of Various Species; (b) Phylogenetic Tree Based on Characteristics

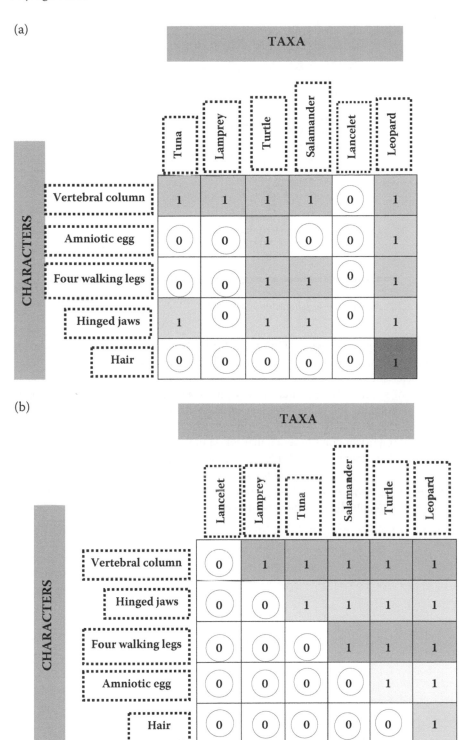

FIGURE 3.21 (a) Character State Matrix; (b) Rearranged Character State Matrix

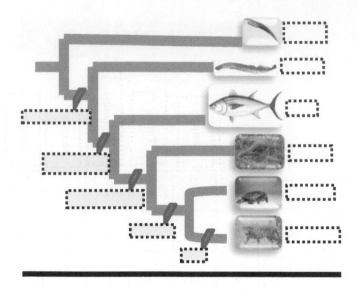

FIGURE 3.22   Cladogram of Character Matrix

phylogenetic tree, as shown in Figure 3.22. On the other hand, a turtle possesses four of the given five characteristics so it will appear closest to the leopard.

Characteristics are actually compared by counting the percentage of similarity in the DNA of species. The resultant phylogenetic tree of Figure 3.23(a) is actually called a cladogram. The actual representation of the phylogenetic tree is shown in Figure 3.23(b).

Another important point to remember while creating cladograms is that one may need to choose among the multiple options to create a cladogram. Several characteristics may have to be considered. The goal here is to have a minimum number of evolutionary steps while moving from one branch to other in a cladogram. An example of a mammal–bird clade and a lizard–bird clade are shown in Figure 3.24.

(a)

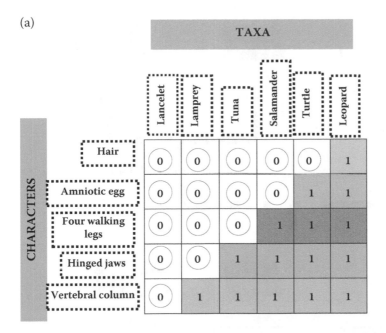

(b)

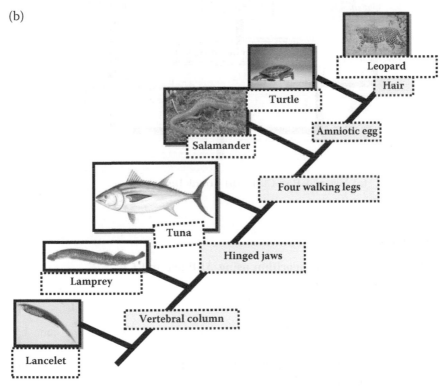

FIGURE 3.23   (a) Character Matrix; (b) Phylogenetic Tree

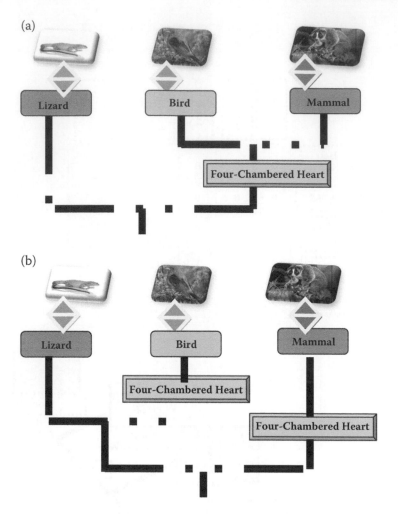

FIGURE 3.24 (a) Mammal–Bird Clade; (b) Lizard–Bird Clade

APPENDIX

Implementing UPGMA algorithm using ASP.net and C#
Distance-Based (UPGMA)

Evolution of new organisms is driven by

- Diversity: Different individuals carry different variants of the same basic blueprint
- Mutations: The DNA sequence can be changed due to single base changes, deletion/insertion of DNA segments, etc.
- Selection bias

Modern biological methods allow the use of molecular features, namely gene sequences and protein sequences. The goal is to generate a tree in which similar sequences with short distances are closer and the sum of the branch lengths of two nodes is equal to their distance.

UPGMA: unweighted pair group method with arithmetic mean.

- Assume a molecular clock (constant evolution rate)
- Produce a rooted tree
- Ultrametric condition: for any three taxa

(a,b,c), dac <= max(dab, dbc).
In other words: the two greatest distances must be equal.
Or: constant evolutionary rate for all branches.

UPGMA condition

$$dAB <= max(dAC, dBC)$$
$$dAC <= max(dAB, dBC)$$
$$dBC <= max(dAB, dAC)$$

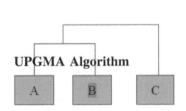

UPGMA Algorithm

Initialization: Define T to be the set of leaf nodes, one for each sequence. The height of each node is zero. Let L = T.

Repeat

- Select the closest two nodes (A, B) and create a parent node K for them. Join A, B and K, respectively. Set the height of node K to dAB/2. Set the branch length between K and A = height K – height A, and set the branch length between K and B = height K – height B.
- Remove A, B from L and add K into L. Re-compute the distance between K and other nodes in L. The distance between K and the other nodes is the average distance of leaf sequences below K and the other node.

Until there is only one node
Design front end

1. Add a text box and button to get the input for the matrix size.
2. Add a DataList to get input matrix coordinates.
3. Add a Div to show output.

C# code explained

1. Generate a Datatable and bind datasource of Datalist to get the input of the matrix elements.
2. Store the Value of the datalist into the datatable (**dtMain**).
3. Open a for loop for the steps.
4. Find the **Min value** and **coordinates** of that min value.
5. Show them in a Label and add the Label to Div to show the step output.
6. Now reconstruct the matrix with new elements. Merge the element's corresponding coordinates of min value. Compute the mean of all the corresponding distances.
7. Now bind the datatable to the gridview and add the gridview to Div to show the step output.
8. Loops execute as per size of the matrix.
9. The output of each step will be available on the aspx page.

Code of ASP.CS page

```
using System;
using System.Collections.Generic;
using System.Data;
using System.Web;
using System.Web.UI;
using System.Web.UI.WebControls;
public partial class upgma: System.Web.UI.Page
 {
 static int step = 0;
 DataTable dtMain = new DataTable();
 protected void btnStart_Click(object sender, EventArgs e)
 {
 int size = Convert.ToInt32(txtSize.Text);
 dtMain = new DataTable();
 for (int i = 0; i <= size; i++)
  dtMain.Columns.Add();
 dtMain.AcceptChanges();
 for (int i = 0; i <= size; i++)
 {
 dtMain.Rows.Add();
 for (int j = 0; j <= size; j++)
 {
  TextBox txt = (TextBox)dlInput.Items[(size * j) + i + j].FindControl("txtInput");
dtMain.Rows[i][j] = txt.Text;
 }
 }
 dtMain.AcceptChanges();
 GridView gvInput = new GridView();
 gvInput.ID = "gvMatrix";
```

```
gvInput.ShowHeader = false;
gvInput.DataSource = dtMain;
gvInput.DataBind();
divResults.Controls.Add(gvInput);
step = size;
DataTable dtTemp = new DataTable();
dtTemp = dtMain;
int r=0, c = 0;
for (int l = 0; l < size-1; l++)
{
decimal min = Convert.ToDecimal(dtTemp.Rows[1][2]);
r = 1;
c = 2;
for (int i = 1; i <= step; i++)
 for (int j = i + 1; j <= step; j++)
 {
  if (min > Convert.ToDecimal(dtTemp.Rows[i][j]))
  {
   min = Convert.ToDecimal(dtTemp.Rows[i][j]);
   r = i;
   c = j;
  }
 }
Label lbl = new Label();
lbl.ID = "lblSteps" + l;
lbl.Text = "Step " + (l + 1);
lbl.Text += "<br/> Min Value is " + min.ToString() + " at coordinates["+ r +","+
c +"]";
lbl.Text += "<br/> Merge the elements of rows and columns corresponding of
(" + dtTemp.Rows[0][c] + ") and (" + dtTemp.Rows[r][0] + ")";
divResults.Controls.Add(lbl);
DataTable dtT = new DataTable();
step--;
for (int i = 0; i <= step; i++)
 dtT.Columns.Add();
int pc = 0;
int pr = 0;
for (int i = 0; i <= step; i++)
{
dtT.Rows.Add();
 for (int j = 0; j <= step; j++)
 {
  if ((i == r || j == r) && (i!= j))
  {
   if ((i == 0 || j == 0))
   {
```

```
dtT.Rows[i][j] = dtTemp.Rows[0][r].ToString() + "," + dtTemp.Rows[0]
[c].ToString();
if (i == 0)
pc++;//increase column index bcoz of column merge from beginning or middle
else
pr++;// increase row index boz of row merge from beginning or middle
}
else if (j > i)
{
dtT.Rows[i][j] = getMeanDistance(dtT.Rows[i][0].ToString(), dtT.Rows[0]
[j].ToString()).ToString();
//dtT.Rows[i][j] = "rc";
}
}
else
{
if (i!= j)
{
if (c!= step+1)
{
if (j > r && i > r)
dtT.Rows[i][j] = dtTemp.Rows[i + pr][j + pc];
else if (j > r)
dtT.Rows[i][j] = dtTemp.Rows[i][j + pc];
else if (i > r)
dtT.Rows[i][j] = dtTemp.Rows[i + pr][j];
else
dtT.Rows[i][j] = dtTemp.Rows[i][j];
}
else
dtT.Rows[i][j] = dtTemp.Rows[i][j];
}
}
}
}
dtT.AcceptChanges();
dtTemp = dtT;
GridView gv = new GridView();
gv.ID = "gvMatrix" + l.ToString();
gv.ShowHeader = false;
gv.DataSource = dtT;
gv.DataBind();
divResults.Controls.Add(gv);
}
}
protected decimal getMeanDistance(string a, string b)
```

```
 {
  decimal rslt = 0;
  string[] stra = a.Split(',');
  string[] strb = b.Split(',');
  for (int i = 0; i < stra.Length; i++)
   for (int j = 0; j < strb.Length; j++)
   {
    int posx = GetCharPosition(stra[i], "From");
    int posy = GetCharPosition(strb[j], "To");
    if (posx < posy)
     rslt += Convert.ToDecimal(dtMain.Rows[posx][posy]);
    else
     rslt += Convert.ToDecimal(dtMain.Rows[posy][posx]);
   }
  rslt = rslt / (stra.Length * strb.Length);
  return Math.Round(rslt,2);
 }
 protected int GetCharPosition(string strChar,string strType)
 {
 int pos = 0;
 if (strType == "From")
 {
  for (int i = 1; i < dtMain.Rows.Count; i++)
   if (dtMain.Rows[i][0].ToString() == strChar)
   {
    pos = i;
    break;
   }
 }
 else
 {
 for (int j = 1; j < dtMain.Columns.Count; j++)
  if (dtMain.Rows[0][j].ToString() == strChar)
  {
   pos = j;
   break;
  }
 }
 return pos;
 }
 protected void btnGenerateMatrix_Click(object sender, EventArgs e)
 {
 DataTable dt = new DataTable();
 int size = Convert.ToInt32(txtSize.Text) + 1;
  dt.Columns.Add();
  for (int i = 0; i < size * size; i++)
```

```
   dt.Rows.Add();
   dt.AcceptChanges();
   dlInput.RepeatColumns = size;
   dlInput.DataSource = dt;
   dlInput.DataBind();
   }
}
```

Output screenshot and execution

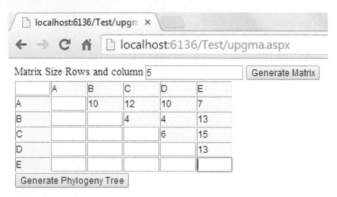

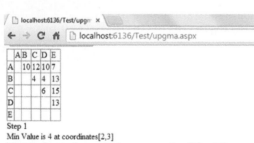

Step 1
Min Value is 4 at coordinates[2,3]
Merge the elements of rows and columns corresponding of (C) and (B)

	A	B,C	D	E
A		11	10	7
B,C			5	14
D				13
E				

Step 2
Min Value is 5 at coordinates[2,3]
Merge the elements of rows and columns corresponding of (D) and (B,C)

	A	B,C,D	E
A		10.67	7
B,C,D			13.67
E			

Step 3
Min Value is 7 at coordinates[1,3]
Merge the elements of rows and columns corresponding of (E) and (A)

	A,E	B,C,D
A,E		12.17
B,C,D		

Step 4
Min Value is 12.17 at coordinates[1,2]
Merge the elements of rows and columns corresponding of (B,C,D) and (A,E)

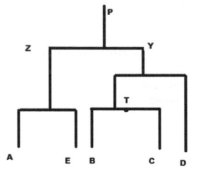

Exercise
  Multiple-Choice Questions

1. A phylogenetic tree that is "rooted" is one

   a. that extends back to the origin of life on Earth.
   b. at whose base is located the common ancestor of all taxa depicted on that tree.
   c. that illustrates the rampant gene swapping that occurred early in life's history.
   d. that indicates our uncertainty about the evolutionary relationships of the taxa depicted on the tree.
   e. with very few branch points.

2. Dozens of potato varieties exist, differing from each other in tuber size, skin colour, flesh colour and shape. One might construct a classification of potatoes based on these morphological traits. Which of these criticisms of such a classification scheme is most likely to come from an adherent of the phylocode method of classification?

   a. Flesh color, rather than skin colour, is a valid trait to use for classification because it is less susceptible to change with the age of the tuber.
   b. Flower colour is a better classification criterion, because below-ground tubers can be influenced by minerals in the soil as much as by their genes.
   c. A more useful classification would codify potatoes based on the texture and flavour of their flesh, because this is what humans are concerned with.
   d. The most accurate phylogenetic code is that of Linnaeus. Classify potatoes based on Linnaean principles, not according to their colour.
   e. The only biologically valid classification of potato varieties is one that accurately reflects their genetic and evolutionary relatedness.

3. Which of the following traits do not help distinguish animals from other forms of life?

   a. the presence of DNA in the cell nucleus
   b. the presence of two types of tissues: nervous tissues for impulse conduction and muscle tissue for movement
   c. cell walls that have structural support
   d. b and c
   e. a and c

4. A cladogram or phylogenetic tree

   a. is a hypothesis about the evolutionary relationships among a group of animal taxa.
   b. is a diagram in which the sequence of branching illustrates the historical chronology of an evolutionary event.

   c. reflects the hierarchical classification of taxonomic groups nested within more inclusive groups.

   d. All of the above.

5. Which describes a phylogeny?

   a. a genealogy
   b. a tree
   c. evolutionary pathways
   d. natural relationships between organisms
   e. All of the above.

6. Species evolve by

   a. diversification.
   b. progression.
   c. linear advancement.
   d. magic.
   e. none of the above.

7. Which if the following is not an algorithm for generating phylogenetic trees from molecular data?

   a. neighbor joining
   b. parsimony
   c. maximum likelihood
   d. Jukes and Cantor
   e. All of the above.

8. The principle of parsimony as used in phylogeny suggests that the simplest tree that fits the data is preferred.

   a. True
   b. False

9. A researcher wants to determine the genetic relatedness of several breeds of dog (*Canis familiaris*). The researcher should compare homologous sequences of _____ that are known to be _____.

   a. carbohydrates; poorly conserved
   b. fatty acids; highly conserved
   c. lipids; poorly conserved
   d. proteins or nucleic acids; poorly conserved
   e. amino acids; highly conserved

10. The most important feature that permits a gene to act as a molecular clock is
   a. having a large number of base pairs.
   b. having a larger proportion of exonic DNA than of intronic DNA.
   c. having a reliable average rate of mutation.
   d. its recent origin by a gene-duplication event.
   e. being acted upon by natural selection.

Essay-Type Questions

Q1.

   a. Discuss the importance of molecular phylogenetics. How are phylogenetic trees useful in retrieving evolutionary information?
   b. Construct a phylogenetic tree using the UPGMA method for the below-mentioned molecular data table.

	A	B	C	D	E
A	0	5	4	9	3
B		0	2	4	3
C			0	5	7
D				0	8
E					0

Q2. What is a clade?

Q3. Draw a phylogenetic tree for the given sequences.

	SEQ1	SEQ2	SEQ3	SEQ4	SEQ5
SEQ1	0	48	51	51	70
SEQ2		0	66	66	85
SEQ3			0	45	80
SEQ4				0	93
SEQ5					0

Q4. Discuss any general technique used to investigate issues like the origin of modern humans and the date of the human/chimpanzee divergence.

4 RNA

Biologically, living organisms must follow certain rules, regulations and guidelines such that the living cell are functioning properly. This very information is carried out by RNA, which stands for ribonucleic acid. It is a universal fact that proteins are building blocks of life. Without the above-mentioned information, by being transferred properly from nucleus to ribosomes, production of accurate proteins is impossible. This very fact conveys the importance of RNA. Underproduction, overproduction or faulty production of proteins leads to diseases.

The structure of RNA is as follows: RNA is single-stranded molecule that folds upon itself to imitate a double helix structure of DNA (dexyribonucleic acid). The backbone of RNA contains ribose as a sugar with four bases: adenine, cytosine, guanine and uracil. Phosphate groups are also attached to ribose. The bases of RNA form base pairs in order to form secondary and tertiary structures. RNA may be synthesized from DNA. DNA (unlike RNA) has a double helix structure and stores the genetic code for long duration. Compared to DNA, RNA is more versatile. Apart from having a significant role in protein synthesis, RNA carries out bio-chemical reactions, gene regulation, methylation, etc. Also, RNA does not have a replicating capability like DNA. The life period of various types of RNA varies. Some of them have a long life and some get destroyed soon after their creation.

4.1 STRUCTURE OF RNA

RNA structure changes from primary structure to secondary structure and on to tertiary structure. All three types of structures are explained below.

4.1.1 PRIMARY STRUCTURE OF RNA

The primary structure of RNA is a sequence of nucleotides (A, G, C and U) that are linked together by phosphodiester bonds.

RNA consists of nucleotides and a backbone. The backbone of RNA is made of a sugar phosphate. The RNA structure has three components: ribose (sugar), phosphate (PO_4) and a nucleotide base. Figure 4.1 shows four bases of RNA, namely adenine, guanine, cytosine and uracil. Figure 4.2 elaborates the structures of the ribosome and phosphate. The chemical structure of RNA is shown in Figure 4.3.

4.1.2 SECONDARY STRUCTURE OF RNA

The bases in a primary structure of RNA form base pairs to conceive a secondary structure. The secondary structure of RNA is characterized with Watson-Crick base pairs (GC or AU) and Wobble base pairs (G-U). GU and AU pairs have two

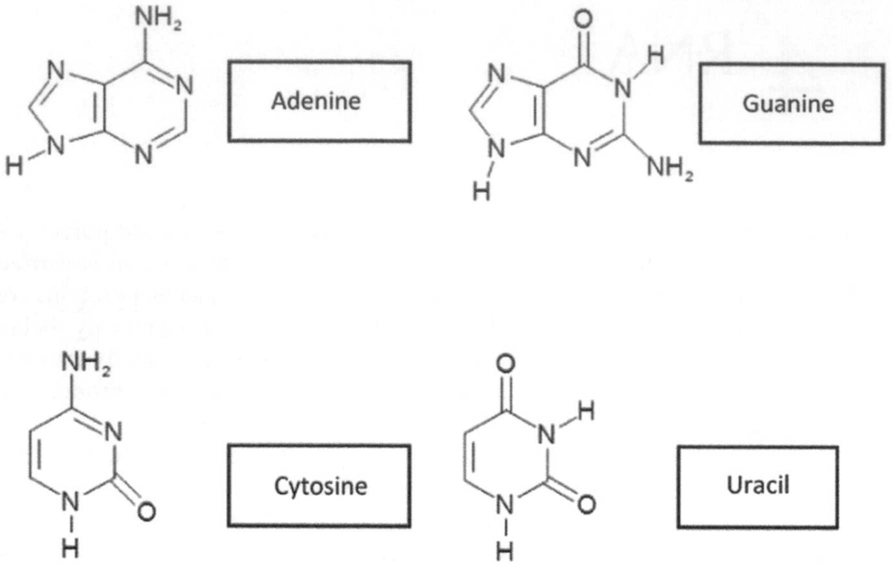

FIGURE 4.1   Four Bases of RNA – Adenine, Guanine, Cytosine and Uracil

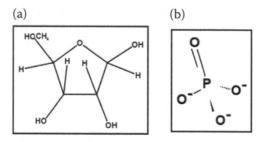

FIGURE 4.2   (a) Ribosome; (b) Phosphate

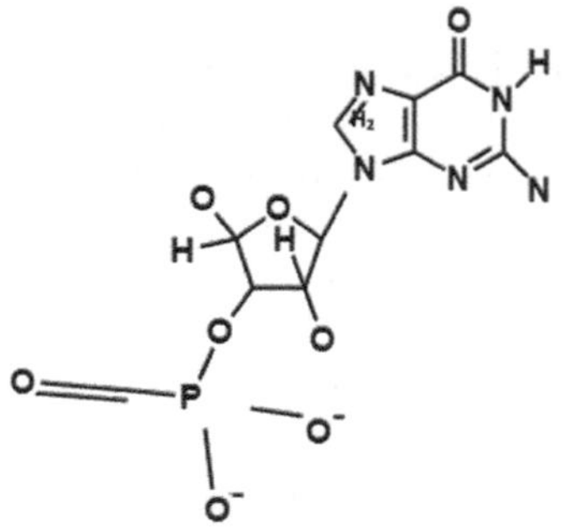

FIGURE 4.3   RNA Chemical Structure

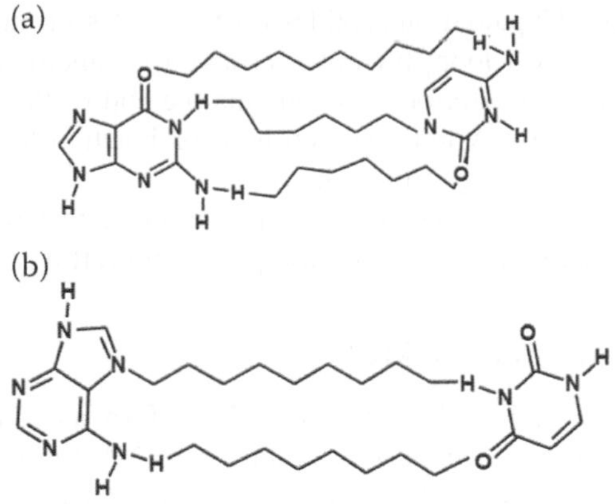

(a)

(b)

FIGURE 4.4 (a) GC Pair; (b) AU Pair

hydrogen bonds. The GC pair has three hydrogen bonds. The GC base-pair structure and AU base-pair structure are shown in Figure 4.4(a) and (b), respectively. Apart from these common base pairs, RNA can even have AC, UU and GA base pairs.

The secondary structures of RNA form specific structures like a hairpin loop, bulge loop, internal loop, exterior loop, pseudoknot, etc., as shown in Figure 4.5. Each structure has its own characteristics and importance.

- Helix   A helix is formed by stacking base pairs. A hairpin loop is shaped when at least three unpaired nucleotides exist in a helix. RNA forms a double helical structure by folding upon itself.
- Bulge Loop – An unpaired nucleotide on either side of a helix is called a bulge loop. It can be termed an asymmetrical structure.

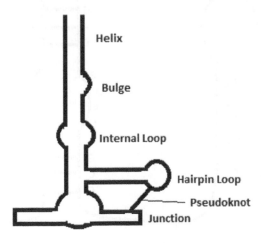

FIGURE 4.5   RNA Secondary Structure Motifs

- Internal Loop – Unpaired nucleotides on both sides of the helix lead to formation of an internal loop. It can be termed a symmetrical structure.
- Exterior Loop – An exterior loop contains the end of the sequence.
- Multi-branch Loop – A multi-branch loop or junction loop is formed when various structures emerge from a helix.
- Pseudoknot – A pseudoknot is a special structure that originates when the above-mentioned structures form base pairs with leftover unpaired bases.

4.1.3 TERTIARY STRUCTURE OF RNA

A secondary structure can form base pairs in order to form a tertiary structure. Some of the unpaired bases in a secondary structure of RNA further form base pairs, resulting in a tertiary structure. A tertiary structure of RNA is common in tRNA.

- Coaxial Stacking of Stems – Long duplexes of RNA stack upon each other to form a long chain. As the length of the chain increases, the chances of forming tertiary interactions also increase.
- Tetraloop Receptor Motif – A common tetraloop nucleotide sequence is GAAA. And then there may be a corresponding receptor, for example CCUAAGUAUGG. A tetraloop receptor is shown in Figure 4.6.

4.2 TYPES OF RNA

RNA has been categorized into three categories according to their nature, function and structure; namely, rRNA, tRNA and mRNA. They play a major role in the

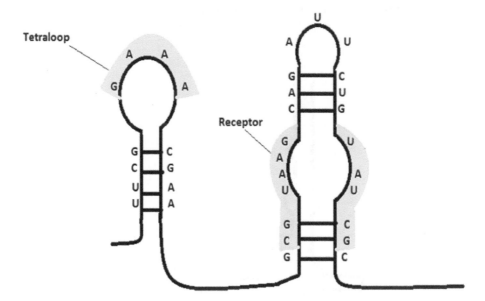

FIGURE 4.6   Tetraloop Receptor Motif

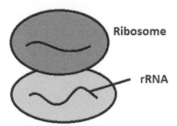

FIGURE 4.7   rRNA

protein assembly process. Apart from these, other types of RNA do exist, with significant roles, as discussed below.

A. *Ribosomal RNA (rRNA):* One of the RNA components of the ribosomes and necessary for protein synthesis. Due to critical functions of rRNA, protein synthesis occurs. A ribosome exists in all organisms and helps translate the information in mRNA (messenger RNA) into a protein. Ribosomes have 60% of rRNA approximately. An rRNA structure is represented in Figure 4.7.

B. *Transfer RNA (tRNA):* Transfer RNA acts as a physical connection between the DNA and RNA sequence of nucleic acids and protein sequence of amino acid. It helps in decoding mRNA (messenger RNA). tRNA is an important component of protein translation and contains 75 to 95 nucleotides. A tRNA structure is represented in Figure 4.8.

C. *Messenger RNA (mRNA):* mRNA is responsible for the transfer of genetic information from DNA to the ribosome and is the biggest family of RNA molecules. mRNA was discovered by two scientists, Elliot Volkin and Lazarus Astachan, in 1956. DNA is copied into mRNA and then decoded into

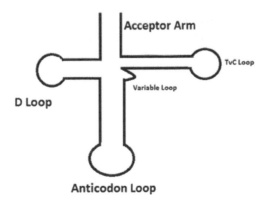

FIGURE 4.8   tRNA

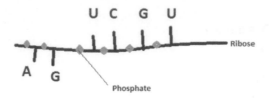

FIGURE 4.9   mRNA

proteins. One molecule of mRNA is needed to encode the information of one protein but for bacteria more than one protein's information can be encoded into one molecule of mRNA. An mRNA structure is represented in Figure 4.9.

D. *Non-coding RNA (ncRNA):* These are functional molecules of RNA that do not encode into protein. However, non-coding RNA contains crucial information and has several functions. One of the functions of ncRNA is to regulate the gene expression at the transcription level.

E. *Transfer-messenger RNA (tmRNA):* These are the RNA molecules that have similar properties like both tRNA and mRNA. Mostly bacteria produce single tmRNA; however, a few permuted tmRNA encode a two-piece tmRNA. It works like a quality control system that monitors the protein synthesis.

F. *Small-nuclear RNA (snRNA):* These are small RNA molecules and are involved in splicing or other reactions of RNA. The approximate length of snRNA is 150 nucleotides. Complexes of snRNA are also known as snRNP (small nuclear ribonucleoproteins), generally called "snurps".

G. *MicroRNA (miRNA):* MicroRNA also comes from a family of non-coding RNA. It plays an important role in regulation of gene expression and RNA silencing. These are single-stranded RNA molecules and are very short in length, only twenty-two nucleotides, approximately. Both animals and plants have MicroRNA.

H. *Small interfering RNA (siRNA):* Small interfering RNA is also known as silencing RNA or short interfering RNA. siRNA is formed from a double-stranded RNA molecular sequence of 20 to 25 base pairs. siRNAs is used for knockdown of long noncoding RNAs genes.

I. *Small nucleolar RNA (snoRNA):* These are a special category of small RNA molecules that signify the chemical changes of other RNAs, mainly ribosomal RNA, transfer RNA and snRNA small nuclear RNAs. It is also proposed that these came from the evolution in gene duplication of transfer RNA (tRNA).

J. *Antisense RNA (asRNA):* These are single-stranded RNA, complementary to messenger RNA. asRNA is sometimes referred to as mRNA-interfering complementary (micRNA) but micRNA is not so popular and not adopted widely. "Sense" is a nucleotide sequence and also results in a protein (a gene product).

K. *Signal recognition particle RNA (SRP RNA):* SRP is a RNP (ubiquitous) complex that results in secretory proteins and membrane. These are required for co-translational protein targeting. The common family known as Alu comes from a gene of 7SL RNA when the central sequence is deleted.

L. *hnRNA (heteronuclear RNA):* Transcription, as described in an earlier chapter, is the process of conception of mRNA from DNA. But this is not actually a one-step process. The intermediate step is conception of pre-mRNA or hnRNA. In other words, it is the first copy of DNA. All introns from hnRNA are removed before completion of the process of transcription. This process is called splicing. hnRNA is a short-lived RNA.

M. *Catalytic RNA or ribozymes:* Catalytic RNA catalyzes chemical reactions in the absence of protein. Self-splicing introns are also catalytic RNA. Ribozymes include peptidyl tranferase, hammerhead, splicesome and hepatitis delta virus (HDV).

N. *Telomerase RNA:* Telomerase RNA molecule is the one that facilitates the addition of a specific set of nucleotides to the 3' end of the duplicating strand for maintaining the telomere end such that the chromosome doesn't fuse with other chromosomes. RNA telomerase is important for the stability of genomes.

O. *gRNA (guide RNA):* During the process of RNA editing, gRNA is used for insertion and deletion of nucleotides in mRNA.

4.3   FUNCTIONS OF RNA

Various functions of RNA are discussed below:

1. The central dogma of molecular biology suggests that progression of information within a cell if from DNA to proteins.
2. The human body needs proteins for a number of functions. The processes of transcription and translation lead to synthesis of proteins. RNA plays a vital role in both of these processes.
3. RNA allows the cell to create amino acids that support gene expression.
4. Through RNA, cells are able to access the information stored in DNA.
5. RNA acts as an enzyme. An enzyme is a substance that speeds up reactions in our body. As RNA is single stranded, it can fold upon itself to change its shape. This is the reason why RNA can act as an enzyme and DNA, with a double strand, cannot.
6. Literature reveals that in the early forms of life, RNA was present instead of DNA.
7. DNA containing the important genetic information has to stay inside the nucleus only. It may be damaged if it moves out, so a copy of part of the DNA is created in the form of mRNA.
8. RNA is prone to alkaline hydrolysis. On passing an alkaline solution, RNA can be degraded without affecting DNA.
9. It is well known that DNA contains hereditary information, but some viruses do not contain DNA. In such cases, RNA holds the genetic information.
10. RNA can function as a catalyst of biochemical reactions, an adapter molecule in protein synthesis, and a structural molecule in cellular organelles.

11. Non-coding RNAs can upregulate or downregulate the expression of genes.
12. RNA regulates the activity of genes under a changing environment or cellular differentiation.
13. Some RNAs, like self-splicing RNA transcripts, ribozymes and RNAse P, acquire intrinsic enzymatic activity and can directly catalyze RNA modification reactions.
14. Riboswitches are a kind of auto-regulatory unit present in mRNA that regulate and are capable of binding to tiny molecules in order to change their structure of the ribosome. They can also turn ON and turn OFF a particular gene.
15. miRNAs impact the growth and functioning of multicellular eukaryotes.
16. MicroRNA (miRNA) leads to gene silencing and degradation.
17. The structural flexibility of RNA leads to the formation of secondary structures which in turn have specific functions.
18. Stem loop structures in RNA can act as binding sites for other regulatory molecules.
19. Small non-coding RNAs can defend the cell against a viral infection.
20. RNA helps the ribosomes choose the right amino acid, which is required in building new proteins in the body.
21. Small nuclear RNA (snRNA) is involved in a splicing activity that leads to production of various proteins from one gene.
22. DNA is useless without RNA.
23. rRNA plays the structural role of RNA wherein it governs the formation of peptide bonds among amino acids.
24. The catalytic role of RNA in ribosomes is played by ribozymes that catalyze some biochemical reactions.
25. The regulatory role of RNA is taken care of by non-coding RNA (ncRNA).

APPENDIX

Exercise

Multiple-Choice Questions

Q1. Which of the following types of RNA code for a protein?

    a. dsRNA
    b. mRNA
    c. rRNA
    d. tRNA

Q2. A nucleic acid is purified from a mixture. The molecules are relatively small, contain uracil and most are covalently bound to an amino acid. Which of the following was purified?

    a. DNA
    b. mRNA

c. rRNA

d. tRNA

Q3. Which of the following type of RNA is known for its catalytic abilities?

a. dsRNA

b. mRNA

c. rRNA

d. tRNA

Q4. Ribosomes are composed of rRNA and what other component?

a. protein

b. polypeptides

c. DNA

d. mRNA

Q5. Which of the following may use RNA as its genome?

a. a bacterium

b. an archaeon

c. a virus

d. a eukaryote

Q6. An unpaired nucleotide on one side of the bulge is called

a. Rnase P.

b. an internal loop

c. a bulge.

d. a junction.

Q7. Ribosomes are composed mostly of RNA.

a. True

b. False

Q8. Double-stranded RNA is commonly found inside cells.

a. True

b. False

Q9. The space between a three-way pairing is called

a. a bulge.

b. a junction.

c. an RNA function.

d. a viroid.

Q10. RNA contains genetic material

a. along with DNA.

b. only when DNA is absent.

c. in both of the above cases.

d. None of these.

Q11. In terms of DNA and RNA structure, what is a nucleotide?

a. A nucleotide is a heterocyclic base.

b. A nucleotide is a sugar molecule covalently bonded to a heterocyclic base.

c. A nucleotide is a sugar molecule bonded to a phosphate group(s) and a heterocyclic base.

d. A nucleotide is a heterocyclic base bonded to a phosphate group(s).

Q12. DNA exists in a double-stranded form, whereas RNA is mainly a single-stranded molecule. What is the likely reason for DNA being double stranded?

a. RNA strands cannot form base pairs.

b. Double-stranded DNA is a more stable structure.

c. DNA cannot exist in the single-stranded form.

d. It is easier to replicate double-stranded DNA than single-stranded RNA.

Q13. What term is used to describe the process by which a segment of DNA is copied to produce a molecule of messenger RNA?

a. reproduction

b. replication

c. translation

d. transcription

Q14. What term is used to describe the process by which proteins are synthesised from a genetic code?

a. reproduction

b. replication

c. translation

d. transcription

Q15. What role does messenger RNA play in the synthesis of proteins?

a. It catalyses the process.

b. It provides the genetic blueprint for the protein.

c. It translates the genetic code to a specific amino acid.

d. It modifies messenger RNA molecules prior to protein synthesis.

Q16. The bases of RNA are the same as those of DNA, with the exception that RNA contains

a. cysteine instead of cytosine.

b. uracil instead of thymine.

c. cytosine instead of guanine.

d. uracil instead of adenine.

Q17. Which one of the following is not a type of RNA?

a. nRNA (nuclear RNA)

b. mRNA (messenger RNA)

c. rRNA (ribosomal RNA)

d. tRNA (transfer RNA)

Q18. Which of the following is the smallest of the RNAs?

a. messenger RNA

b. transfer RNA

c. ribosomal RNA

d. All of these.

Q19. What is the function of messenger RNA?

a. It carries amino acids.

b. It is a component of the ribosomes.

c. It is a direct copy of a gene.

d. It is the genetic material of some organisms.

Q20. The ratio of mRNA to total RNA in cell is

a. 3–4%.

b. 3–6%.

c. 3–8%.

d. 3%.

Essay-Type Questions

Q1. What are the differences between DNA and RNA?

Q2. What does RNA stand for?

Q3. Discuss various types of RNA.

Q4. What are the major functions of RNA?

Q5. Can RNA contain genetic material just like DNA? Support your answer.

Q6. Discuss the chemical structure of RNA with a suitable example.

5 Pseudoknot

The four bases in a primary sequence of RNA are adenine (A), guanine (G), cytosine (C) and uracil (U) which form a base pair. In other words, the primary sequence of RNA folds upon itself to form a secondary structure. The secondary structure in turn folds to form a tertiary structure. The function of RNA depends on its structure. It is quite costly to predict the tertiary structure of RNA. So, the secondary structure of RNA is focused upon by researchers for its accurate prediction. Various motifs in RNA secondary structure have been identified such as hairpin loop, bulge loop, internal loop, multi-branch loop, etc., as mentioned in Chapter 4.

Another important motif that was ignored by researchers for many years while predicting the RNA secondary structure is the pseudoknot. Reasons for ignoring the pseudoknot:

- The presence of the pseudoknot makes the structure complex.
- Prediction of the optimal structure using the ab initio recursive algorithms is not possible.
- The frequency of occurrence of pseudoknots is less as compared to other motifs.

Nevertheless, the prediction accuracy of RNA secondary structure prediction algorithms cannot be improved if pseudoknots are ignored. The major structural difference between a pseudoknot and other motifs is the way bases are paired, as shown in Figure 5.1(a) and (b). Without pseudoknot, the structure is recursive but the presence of a pseudoknot breaks the recurrence relation, thereby making its prediction difficult.

5.1 DEFINITION

The single-stranded RNA folds upon itself to form a secondary structure. A pseudoknot results when a loop in that secondary structure pairs with a complementary sequence outside the loop, as shown in Figure 5.1(a) and (b).

A pseudoknot is just two overlapping base pairs. Two base pairs, i, j and i', j', respectively, are overlapping if $i < i' < j < j'$ or $i' < i < j' < j$. So, base pairing in a pseudoknot occurs in a non-nested fashion, contrary to the base paring in a non-pseudoknotted structure.

5.1.1 PSEUDOKNOT VS KNOT

A pseudoknot is sometimes called a false knot. If two ends of a pseudoknot are pulled in opposite directions, it turns into a straight line. This is the reason why pseudoknot is termed a false knot. Contrary to it, if two ends of a true knot are

(a)   (b)

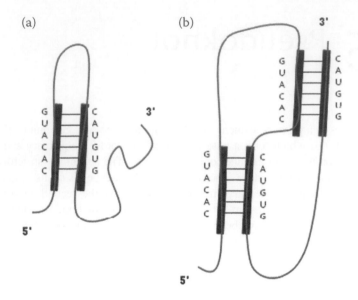

FIGURE 5.1 (a) Non-Pseudoknotted Structure; (b) Pseudoknotted Structure

stretched in the opposite direction, it tightens the knot instead of releasing it. Recently, in 2015, such a knot had been observed in an umbilical cord with an abnormally large length, as shown in Figure 5.2. Such a knot may be a result of increased foetal movement. Other cases of knots in umbilical cords are shown in

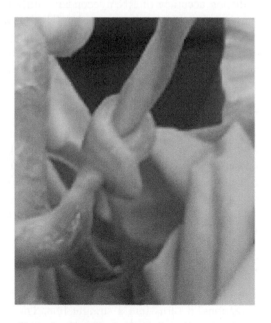

FIGURE 5.2 True Knot in Umbilical Cord (http://www.pathologyoutlines.com/topic/placentaknots.html)

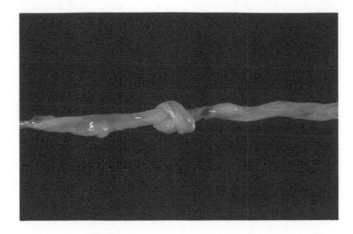

FIGURE 5.3   True Knot in Umbilical Cord (http://library.med.utah.edu/WebPath/PLACHTML/PLAC010.html)

Figures 5.3 and 5.4. The difference in a knot and pseudoknot can be easily spotted in Figures 5.4 and 5.5.

5.1.2   EXAMPLE OF PSEUDOKNOT

Literature shows various examples of pseudoknots, as shown in Figures 5.6 to 5.10. They are:

- a tRNA-like structure of turnip yellow mosaic virus (TYMV)
- a mouse mammary tumor virus (MMTV)
- hepatitis delta virus

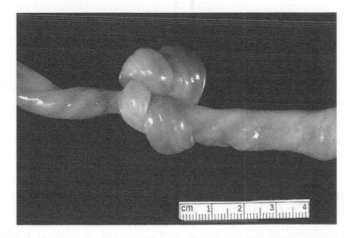

FIGURE 5.4   Knot in Umbilical Cord (http://library.med.utah.edu/WebPath/PLACHTML/PLAC028.html)

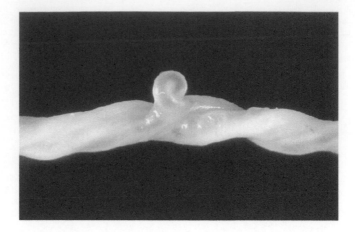

FIGURE 5.5   Pseudoknot in Umbilical Cord (http://library.med.utah.edu/WebPath/ PLACHTML/PLAC073.html)

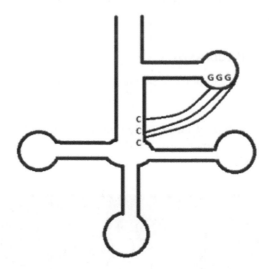

FIGURE 5.6   Pseudoknot – tRNA-like Structure of TYMV RNA

- influenza A
- Potato Yellow Vein Virus (PYVV)

5.2 BIOLOGICAL SIGNIFICANCE OF RNA PSEUDOKNOT

A pseudoknot was first discovered in the turnip yellow mosaic virus in 1982 [10]. Pseudoknots have many varied functions. One of the significant functions of a pseudoknot is frameshifting. It induces ribosomes into an alternative reading frame.

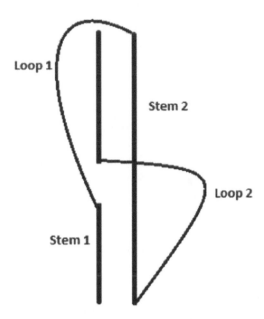

FIGURE 5.7   Pseudoknot – Mouse Mammary Tumor Virus (MMTV)

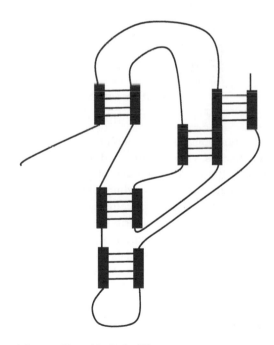

FIGURE 5.8   Pseudoknot – Hepatitis Delta Virus

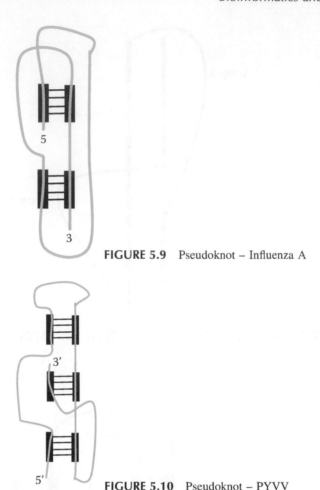

FIGURE 5.9   Pseudoknot – Influenza A

FIGURE 5.10   Pseudoknot – PYVV

Pseudoknots comprise functional domains within ribozymes, self-splicing introns, ribonucleo protein complexes, viral genomes and many other biological systems. It is also required for the activity telomerase that plays an important role in aging. A pseudoknot has a fundamental role in re-coding when it is present with messenger RNA. It also plays regulatory roles in protein synthesis. Pseudoknots have been found in many RNA classes; for example, ribosomal RNA, mRNA, tmRNA, catalytic and self-splicing RNA, viral genomic RNA, telomerase, etc. A pseudoknot leads to translational control, maintains signals for replication and is also involved in catalytic activities. Pseudoknots in a viral RNA role have a crucial role in virus gene expression and genome replication.

5.3   REPRESENTATIONS OF A PSEUDOKNOT

A pseudoknot can be represented in various forms. All the representations have to preserve the crossing of base pairs, as shown in Figure 5.11.

FIGURE 5.11 Base Pairs of RNA, Crossing Each Other

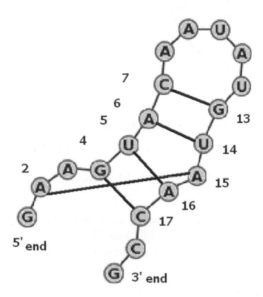

FIGURE 5.12 Planar Graph Representation of a Pseudoknot

5.3.1 PLANAR GRAPH REPRESENTATION

The planar graph structure of a pseudoknot can be seen as a graph, consisting of vertices and edges, as shown in Figure 5.12. The bases of RNA are vertices and the base pairs represent edges. Note that the planar representation of a pseudoknot has crossing edges.

5.3.2 CIRCULAR REPRESENTATION

Now consider the planar structure once again. If the two ends of a planar structure are joined to form a circle, stretching the length of base pairs, the resultant representation is called a circular representation, as shown in Figure 5.13. Crossing of edges in this representation ensures the presence of a pseudoknot in the structure.

5.3.3 DOT BRACKET REPRESENTATION

A dot bracket notation is a one-dimensional representation of RNA structure. All the bases of the RNA sequence are written from the left to right direction, as shown in Figure 5.14. Non-crossing base pairs are represented by round brackets ((,)).

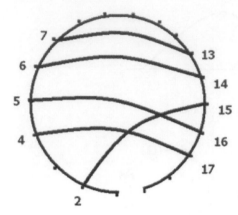

FIGURE 5.13 Circular Representation of Pseudoknot

Sequence: **GAAGUACAAUAUGUAACCG**

Structure: . { . ((((. . . . .)) })) . .

FIGURE 5.14   Dot Bracket Representation of Pseudoknot

Crossing base pairs are then represented by curly brackets ({,}). More crossing base pairs can be represented by square brackets [[,]]. Any opening bracket and its corresponding closing bracket represent a base pair.

5.3.4 Arc Representation

The arc representation of a pseudoknot is also one dimensional, consisting of bases written from the left to right direction. Base pairs are represented by arcs. The number of arcs is equal to the number of base pairs. For a pseudoknot, at least two arcs cross each other, as shown in Figure 5.15.

5.4 TYPES OF PSEUDOKNOTS

Pseudoknots have been categorized into various types, depending upon their base-pairing structure. All of the types are discussed below.

FIGURE 5.15   Arc Representation of Pseudoknot

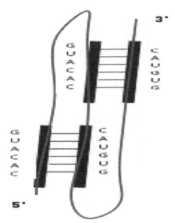

FIGURE 5.16   H-Type Pseudoknot

5.4.1   H-TYPE PSEUDOKNOT

The most common type of pseudoknot is an H-type pseudoknot, shown in Figure 5.16. An H-type pseudoknot is an extension of a hairpin structure of RNA. The hairpin structure forms base pairs with unpaired bases of leftover strands, thereby forming crossings of base pairs. PseudoBase++ [11], an extension to PseudoBase, contains 236 H-type pseudoknots.

5.4.2   H-H TYPE PSEUDOKNOT

An H-H type of pseudoknot includes base pairing between two hairpin loops, as shown in Figure 5.17. PseudoBase contains one sequence with an H-H type pseudoknot.

5.4.3   H-L TYPE PSEUDOKNOT

An H-L type of pseudoknot is formed when base pairing occurs between a hairpin loop and a single-stranded part of a bulge loop or an internal or multiple loop. An H-L type of pseudoknot occurs in two forms: H-L_{out} type pseudoknot and H-L_{in} type pseudoknot. PseudoBase has twenty-four sequences that contain an H-L_{out} type of pseudoknot and eleven sequences that contain an H-L_{in} type of pseudoknot.

5.4.3.1   H-L_{out} Type Pseudoknot

An H-L_{out} type of pseudoknot is shown in Figure 5.18 and is formed when base pairing occurs between a hairpin loop and an internal loop or multiple internal loops.

5.4.3.2   H-L_{in} Type Pseudoknot

An H-L_{in} type of pseudoknot, as shown in Figure 5.19, is formed when base pairing occurs between a hairpin loop and an external loop or multiple external loops.

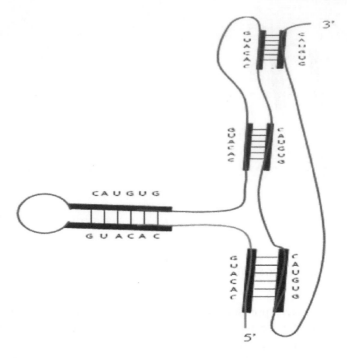

FIGURE 5.17   H-H Type Pseudoknot

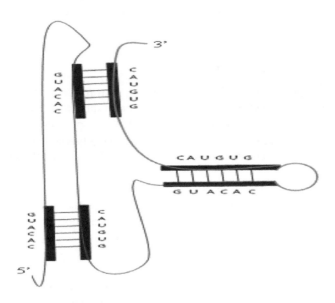

FIGURE 5.18   H-L$_{out}$ Type Pseudoknot

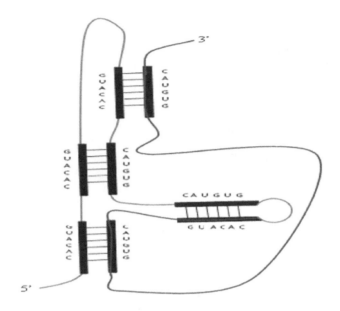

FIGURE 5.19   H-L$_{in}$ Type Pseudoknot

5.4.4   L-L TYPE PSEUDOKNOT

An L-L type of pseudoknot is formed when base pairing occurs between two bulge loops, as shown in Figure 5.20. PseudoBase has thirteen sequences that contain L-L type pseudoknots.

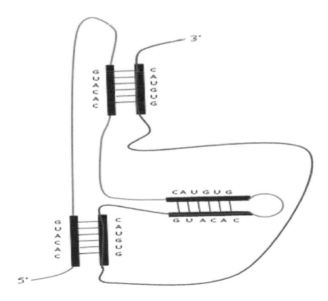

FIGURE 5.20   L-L Type Pseudoknot

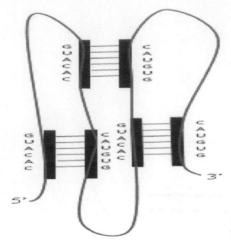

FIGURE 5.21 Kissing Hairpin or H-H-H Type Pseudoknot

5.4.5 KISSING HAIRPIN PSEUDOKNOT OR H-H-H TYPE PSEUDOKNOT

Base pairing between multiple hairpin loops forms an H-H-H type of pseudoknot, as shown in Figure 5.21. Three hairpin loops are formed in sequence, one after the other. Afterwards, these hairpin loops form base pairs among their unpaired bases. Kissing hairpins are essential for replication in the Coxsackie virus. They also play a role in viral plasmid DNA replication or RNA synthesis. This interaction of two loops is exceptionally stable.

5.4.6 THREE-KNOT PSEUDOKNOT

A three-knot pseudoknot is formed by base-pair interactions among three H-type loops, as shown in Figure 5.22.

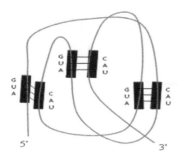

FIGURE 5.22 Three-Knot Pseudoknot

5.4.7 Closed Five-Chain Pseudoknot

A closed five-chain pseudoknot is formed by base-pair interactions among five H-type loops, as shown in Figure 5.23.

5.4.8 Multiloop Pseudoknot

A multiloop pseudoknot is formed when a stem in a secondary structure encloses one or more pseudoknots. A pseudoknot multiloop of tmRNA is shown in Figure 5.24. The substructures P1, P2, P3, P4 and P5 are H-type pseudoknots; Stem1 and Stem2 enclose pseudoknots.

5.4.9 I-Type Pseudoknot

An I-type pseudoknot is the one that involves base pairing between an internal loop and a leftover unpaired single-stranded region of a secondary structure. An I-type pseudoknot is shown in Figure 5.25.

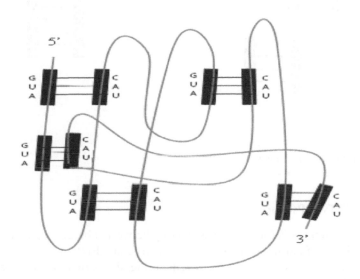

FIGURE 5.23   Closed Five-Chain Pseudoknot

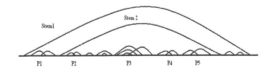

FIGURE 5.24   Multiloop Pseudoknot

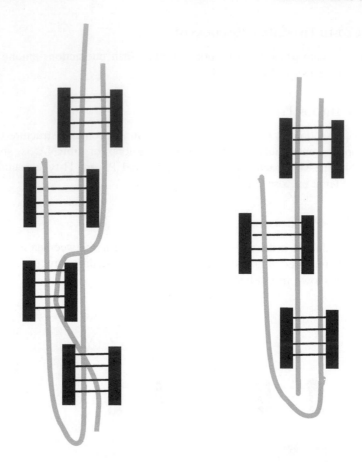

FIGURE 5.25   I-Type Pseudoknot

5.4.10   B-Type Pseudoknot

A B-type pseudoknot is formed by base pairing between a bulge loop and a leftover unpaired single-stranded region of a secondary structure. Both of the bases in such a base pair are actually unpaired bases left during the formation of a secondary structure, as shown in Figure 5.26.

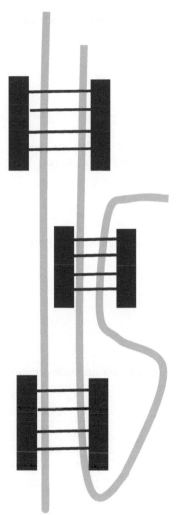

FIGURE 5.26   B-Type Pseudoknot

APPENDIX

Exercise
Multiple-Choice Questions

Q1. Which of the following is not a type of RNA pseudoknot?

    a. H-Type
    b. H-H Type
    c. H-L Type
    d. L-Type

Q2. The _____ refers to the base pairing of a loop in a secondary structure with a leftover complimentary sequence.

    a. hairpin loop
    b. interior loop
    c. pseudoknot loop
    d. helical junction

Q3. Five main subtypes of secondary structures can be identified. They are hairpin loops, bulge loops, interior loops, multi-branch loops and pseudoknots.

    a. True
    b. False

Q4. The _____ refers to a structure with two ends of a single-stranded region (loop) connecting a base-paired region (stem).

    a. helical junction
    b. hairpin loop
    c. bulge loop
    d. interior loop

Q5. The _____ refers to two single-stranded regions on opposite strands connecting two adjacent base-paired segments.

    a. hairpin loop
    b. interior loop
    c. pseudoknot loop
    d. helical junction

Q6. A pseudoknot is different from a knot.

    a. True
    b. False

Q7. A pseudoknot can be represented by

    a. a dot matrix representation.
    b. a circular representation.
    c. both a) and b).
    d. none of these.

Q8. Pseudoknots comprise functional domains within _____ .

    a. ribozymes
    b. self-splicing introns

c. DNA

d. both a) and b)

Q9. An example of a pseudoknot is

a. a mouse mammary tumor virus (MMTV).

b. the hepatitis delta virus.

c. influenza A.

d. PYVV.

e. All of the above.

Q10. PseudoBase is a pseudoknot database.

a. True

b. False

Essay-Type Questions

Q1. What are pseudoknots? Are they real knots? Support your answer with a suitable example.

Q2. Discuss various types of pseudoknots and their base-pair interactions.

Q3. Pseudoknots occur infrequently in RNA, but it is still an important motif in RNA. Support your answer and discuss the importance of pseudoknots.

Q4. What are the representations of a pseudoknot? Explain with a suitable example.

Q5. An H-L type of pseudoknot and an L-L type of pseudoknot appear to be similar. Explain the difference between the two.

6 Pseudoknot Prediction Techniques

Predicting the secondary structure of RNA helps to understand its function. RNA secondary structure has been predicted by various algorithms [13–17]. RNA secondary structure prediction started with the concept of base-pair maximization [18] in which prediction is based on the maximum number of base pairs in the structure. The prediction is based on dynamic programming wherein the given problem is solved by a divide-and-conquer policy. The problem is divided into sub-problems recursively. The sub-problem at the lowest level is then solved. The solutions of sub-problems are used to find the solution of the given problem recursively.

$$S(p, \; r) = \max \begin{cases} S(p, \; r-1) \\ \max \; (S(p, \; q-1) + S \; (q+1, \; r-1) + 1) \\ p < q < r \end{cases}$$

The recurrence relation, as shown above, has been used to solve the RNA secondary structure prediction problem. $S(p, r)$ denotes the RNA secondary structure prediction problem where p and r denotes the first and last nucleotide or base of the RNA sequence to be folded into a secondary structure. The problem $S(p, r)$ is recursively divided into sub-problems $S(p, q-1)$ and $S(q+1, r)$ where q denotes a base that is somewhere between p and r in the RNA sequence. The sub-problem $S(p, r-1)$ denotes that the base r is not paired with any other base in the RNA sequence. In case r is paired, the problem is divided recursively into multiple sub-problems. The final RNA secondary structure is predicted by solving the sub-problems and recursively adding their solutions to build the total solution. Base-pair maximization laid the foundation for another technique called a thermodynamic energy model [12]. In this technique, the estimated energy parameters of sub-structures of RNA secondary structure are added to find the total energy of an entire secondary structure. The secondary structure with a minimum value of thermodynamic energy is considered the optimal secondary structure.

The energy model of [12] was further improved by [18]. Formal grammar techniques were also used to predict RNA secondary structures. An attempt was made in 2003 to predict a pseudoknot using stochastic context-free grammar [19]. This model uses homologous RNA sequences and their phylogenetic trees as input. The model used the CYK algorithm and an inside-outside algorithm to generate the output structures. Their web server is available at www.daimi.au.dk/~compbio/pfold. The model claims to produce high-quality results. The motifs that have been predicted in the secondary structure are hairpin loop, stacked pair, bulge loop, internal loop and junction. Various motifs in RNA are conserved in structure i.e. even if two RNA sequences are different, they may contain the same motif. Like other motifs, a pseudoknot is also a significant motif in an RNA secondary structure. Pseudoknots perform crucial functional roles like alteration of gene expression by inducing ribosomal frame shifting, regulation of translation and splicing, mimicry and selinocystein biosynthesis. So, pseudoknots cannot be ignored for structural and functional analysis of RNA. Also, comparative sequence analysis approach for secondary structure prediction has revealed conserved pseudoknots e.g. in rRNAs, ribonuclease P RNAs and tmRNA. The structural, functional and sequence data on pseudoknots have been gathered in databases like PseudoBase [20], RFAM, etc. [21] Pseudoknot prediction models can be categorized into seven categories on the basis of literature survey. These categories are dynamic programming models, comparative approach-based models, heuristic models, formal grammar models, inverse folding models, integer programming models and soft computing models.

The most popular category model for an RNA secondary structure prediction including pseudoknots is dynamic programming [22–27]. Dynamic programming models require estimation of energy terms contributing to a secondary structure iteratively. Unfortunately, dynamic programming models demand huge time and memory requirements. It is NP-hard to predict RNA secondary structures with arbitrary pseudoknots by free energy minimization. Therefore, some models use a modified form of dynamic programming [28,29]. The model proposed by [28] is a modification of dynamic programming for sequence alignment algorithm. The model proposed by [29] is a sparsification algorithm that introduces the concept of sparsification of matrices used in dynamic programming algorithm in order to reduce space and time requirements. The second category of models for a pseudoknot prediction is comparative analysis [30,31]. Comparative models use covariance score based on maximum weighted matching concept. They are reliable methods but their performance depends on accuracy of alignment of homologous sequences. However, they can work well for long sequences. The third category of Pseudoknot prediction models is heuristics [32,33]. Heuristic techniques are applied in an ad hoc manner. These models execute fast as compared to dynamic programming models but they do not guarantee the optimality of structure. The fourth category focuses on a pseudoknot prediction based on formal grammar [34–38]. RNA secondary structure prediction using grammar is viewed as a parsing problem. The fifth category is inverse folding models [39]. An inverse folding model takes as input the target structure and outputs the RNA sequence. The sixth category is integer programming models [40,41]. These models use pre-built integer programming software for carrying dense calculations. For this reason, these algorithms can be applied on longer sequences. Recently, researchers have tried to solve the pseudoknot prediction problem with soft computing techniques like support vector machine [42], neutral networks [43], genetic algorithms [44] and neural networks [45]. These are the techniques that are generally needed to be combined with some other approach like dynamic programming to solve the prediction problem. Various other soft computing techniques to realize an RNA secondary structure have been discussed in [46].

6.1 DYNAMIC PROGRAMMING TECHNIQUE

Dynamic programming models assume a recursive solution for a given problem, sub-solutions being recorded for reuse. The minimum free energy of a structure is recursively calculated by adding the energy of its sub-structures. The energy of a sub-structure is calculated using standard thermodynamic parameters [27]. Unfortunately, the standard thermodynamic information for pseudoknots does not exist. So, researchers rely on its approximate values, leaving room for optimization. As the theory behind dynamic programming model and free energy minimization is "most stable is the most likely", so it computes one optimal structure of minimum energy. One of the earliest attempts to predict pseudoknots in an RNA secondary structure using dynamic programming was made by [22]. For predicting optimal minimum free energy structure for single sequence, they proposed various recurrence relations to find score for various sub-structures like bulges, hairpin loops and multi-loops as well as pseudoknots. Their model (pknots) calculated a minimum free energy score by adding the score of all its sub-structures. The score of non-pseudoknotted substructures was calculated using standard thermodynamic parameters while for pseudoknots, approximated parameters were used. The efficiency of pknots depends on the score parameters. So, it is important to have accurate values for these parameters. The pseudoknot structures were represented by graphical representation using gap matrices and recurrence relations were formed with the help of those gap matrices. This algorithm can handle structures having up to two gap matrices i.e. very complex structures are ignored. While calculating minimum free energy score of sub-structures, this algorithm also considered coaxial stacking energies for nested and non-nested structures with the help of estimated score parameter. This again creates room for optimization. Although this algorithm worked well for small RNA sequences without producing any spurious pseudoknots, the biggest disadvantage of this model is its too high memory and time requirements due to consideration of the broadest class of Pseudoknot structures.

Another researcher [24] used the recurrence relations of [22] and proposed another model based on dynamic programming. But his recurrence relations were more restricted than recurrence relations proposed by [22]. This dynamic programming model is also inspired from a model proposed by [47]. It is a complicated and difficult-to-understand Tree Adjoining Grammar (TAG) approach to predict pseudoknots. Uemura's concept was used by [24] but without TAGs, making the algorithm simple and flexible. It is observed that both algorithms had the same time complexities. Initially, in this algorithm a scoring function was used that was based on the total number of base pairs. Then this algorithm was extended in various ways. Firstly, this algorithm was extended to include energy function based on adjacent base pairs. Another extension was the consideration of free energy of loop regions. Due to the unavailability of thermodynamic information for pseudoknots, the energy of loops was approximated from the length of the loop. This is where the algorithm's performance could be enhanced. The free energy parameters of the RNA secondary structures were being worked upon by [48] including H-type pseudoknots to improve the performance of their model [49]. They have introduced a penalty factor while calculating the total free energy of the predicted structure. For predicting the energy values, sub-sequences of lengths 3, 4 and 5 nucleotides, respectively, are collected. Their occurrence is counted in their training dataset and the count value is then normalized. They claim to predict energy values with more numbers of factors other than the count of occurrence as the accuracy of their model is not 100%. A time approximation algorithm was proposed by [24] with $O(n^4-\delta)$ by applying his previously proposed approximation technique [50] on his dynamic programming model. The term δ was used as any fixed constant from 0 to 1. The basic dynamic programming model of Akutsu worked for simple pseudoknots. So, another extension was to include recursive pseudoknots. The extensions applied to the proposed basic algorithm raised the time complexity to $O(n^5)$ which is impractical. Akutsu proved that pseudoknots can be handled by dynamic programming. But he also provided strong mathematical basis to prove that the pseudoknot prediction of generalized pseudoknots is NP hard.

A model named pknotsRG, proposed by [26], is also based on the model of [22]. They further restricted the class of pseudoknot structures being used. pknotsRG extended the standard folding algorithms by the class of canonical simple recursive pseudoknots. They defined a class of structures (sr-pk) that excluded triple-crossing helices as well as kissing hairpins in order to reduce the number of candidate structures. The class of structures, sr-pk, was further restricted by applying three rules of canonization. The resultant class was named csr-pk. These rules put a few restrictions on helices of a pseudoknot structure such as length of helices, excluding bulges, etc. It computed not only the minimal free energy structure for a sequence but also other k best structures, displayed as a dot bracket string. Three variants of this model have been introduced: pknotsRG-mfe, pknotsRG-enf and pknotsRG-loc. pknotsRG-mfe produced a minimum free energy structure that could contain a pseudoknot occasionally. pknotsRG-enf produces an energetically best structure having at least one pseudoknot. pknotsRG-loc produced a structure with a pseudoknot at a particular location. Unlike other models that implement their algorithm in C, C++ or other programming language, pknotsRG implements their algorithm in algebraic dynamic programming. The motive of using algebraic dynamic programming was to make the implementation reusable. The algorithm executes fast and has low memory requirements as compared to [22] but on the expense of limitations applied to structures. The earlier dynamic programming models ignored an important concept related to folding of RNA from primary sequence to the stable tertiary structure. This concept is termed hierarchical folding and states that the primary structure forms the secondary structure that in turn forms the tertiary structure. In this respect, a pseudoknot free structure is formed first, and then a pseudoknot forming base pairs is added to form a pseudoknotted structure. The HFold model [51] used this concept to predict pseudoknot structures. Their algorithm could take as an input an RNA sequence and output structures that could be divided into two pseudoknot-free structures. They could predict the kissing hairpin pseudoknot as an interaction between two H-type pseudoknots. In 2014, another model named CCJ [52] was proposed to predict kissing hairpins. This model worked on a minimum free energy concept. It predicted a few types of pseudoknots in $O(n^5)$ time. An extension to HFold was proposed in 2014 [53]. Iterative HFold works on the minimum free energy concept with hierarchical folding. It takes as input a pseudoknot free structure and produces a pseudoknotted structure. In iterative HFold models, authors have tried to reduce the space complexity while predicting the same structures as HFold.

Another dynamic programming model was proposed by [27]. This model is an extension to algorithm of [54] to incorporate arbitrary pseudoknots. The model worked with a motive to align arbitrary pseudoknots using the dynamic programming concept while producing optimal structures. Optimal alignment was calculated by a minimum edit distance using operations such as base deletion, base substitution, arc mismatch, arc breaking, arc removing, arc altering etc. The alignment cost was calculated recursively by calculating and adding the cost of these subalignments. RNA structure was classified into two categories: simple structure and complex structure. Accordingly, simple and complex logic was applied respectively on these categories. Because of this reason, the running time could be substantially lowered. Also, it created room for further optimization. The running time and memory requirements of some of the well-known dynamic programming models are also reduced [29]. The method used is sparsification to handle gapped fragments. Sparsification in general reduces the number of candidate substructures that automatically reduces the time requirements as well as memory requirements. It sparsifies the popular algorithms [22–26]. In [22], sparsification is done for base-pair maximization algorithm by providing recursion equations that are close to the actual algorithm's equations. Also, in its free energy minimization algorithm, recursions in matrices are sparsified. In the algorithm proposed by [26], sparsification is achieved by restricting its scoring scheme. Models proposed by [23–25] were considered a restriction to the class of structures of [22]. So their sparsification is done more or less similarly.

Another modification of dynamic programming models was proposed by [28]. The pseudoknot Local Motif Model and Dynamic Partner Sequence Stacking (PLM_DSS) model predicts the RNA pseudoknot with high sensitivity. It is a modification of dynamic programming for sequence alignment algorithm that calculates lowest stacking energy between stem-forming regions. It assumes the neighbouring regions to be interference-free. It outputs all pseudoknots compatible to the PLM model. PLM model is formed by applying various constraints while selecting H-type pseudoknots from PseudoBase. PLMM_DPSS first predicts potential stems from the RNA sequence by applying a pseudoknot stem energy restriction. A stem is selected by calculating the lowest Turner's stacking energy. After stem selection, they are assembled to predict pseudoknots. This algorithm did output not only pseudoknots of lowest energy but also other compatible pseudoknots. Though the algorithm predicted a pseudoknot with high sensitivity, but it does not take into account non-crossing stem regions. Also, it constructed a PLM model by applying various restrictions on pseudoknots like maximum loop size allowed so it can be applied to a very restricted class of structures. The major drawback of dynamic programming models is their time and space demands. Researchers continued to work in this respect. Researcher Wong proposed a memory efficient dynamic programming model [55] based on Maximum Expected Accuracy (MEA). This algorithm worked on the same concept as proposed by [56] in their model PAL but it could do the same work with lower memory requirements and same time requirements. Both of these algorithms structurally align ncRNA by finding a known query sequence including embedded simple pseudoknots in the given sequence with unknown structure. This memory efficient algorithm extends a PAL algorithm by a divide-and-conquer approach in order to produce a maximum structural alignment score. It works only for embedded simple pseudoknots. The concept of structural alignment of [56] was extended [57] to include sequence similarity as well for a RNA secondary structure prediction.

6.2   COMPARATIVE SEQUENCE ANALYSIS TECHNIQUE

Comparative approach-based pseudoknot prediction models are based on the fact that structures conserved by evolution are more likely to be the functional form. Comparative approach-based pseudoknot prediction models require multiple aligned sequences as input. The sequence similarity should be such that they can be initially aligned. Also, they should be dissimilar enough so that conserved residues and covariant base pairs may be found. If the sequences are too different, comparison becomes difficult. So, it is important to understand which homologous sequences should be used. They have suggested a method to choose sequences for comparative analysis based on GC stability and GU intermediate state constraints.

A comparative approach-based pseudoknot prediction can be applied to a bisecondary structure that is defined as a superposition of two disjoint secondary structures. The model Hxmatch [30] is based on alignment of sequences and it assigns score to base pairs on the basis of thermodynamic score and covariance score. Using the notion of bisecondary structure, it converts the structure to bisecondary structure by deleting the conflicting helices. This Maximum Weighted Matching procedure is repeated until all such helices are deleted. In order to perform well, this model requires accurate alignment with a high sequence covariation. It sometimes produces spurious pseudoknots, too. Another model of a probknot was proposed by [31]. Probknot is a model similar to a maximum weight matching concept. It uses base pairs of highest pair probabilities and the structures having a maximal sum of pairing probabilities are termed as maximum expected accuracy structures. These base-pair probabilities are predicted using a partition function that uses coaxial stacking. From base-pairing probabilities, maximum expected accuracy structures are formed. This model predicts many pseudoknots, but its percent of predicted pairs in known structures is low. Also, the partition function used does not explicitly handle pseudoknots. A comparative sequence analysis with homologous sequences produces better prediction accuracy. DotKnot-PW algorithm [58] uses two homologous sequences for prediction of an H-type pseudoknot. An H-type pseudoknot candidate is first produced using DotKnot [59], which is a single sequence prediction method. In order to reduce computational time, a pairwise base similarity score is calculated instead of a full structure alignment. Optimal structures have been output using a combined free energy and similarity score. Recently, a comparison of Ipknot [41], CCJ [52], Iterative HFold [53] and Hotknots [60], is given by [61]. They have compared the models on same-energy parameters. They have shown that none of the models have achieved PPV and sensitivity values are more than 0.9.

6.3   HEURISTICS

Heuristic pseudoknot prediction models are based on experimentation. These models are designed to reduce the computational time of approaches like dynamic programming. An earlier attempt in heuristic-based pseudoknot predictions was made by [62]. This model (Kinefold) predicted pseudoknots by combining simple structural models and exact clustered stochastic (ECS) simulation. The ECS simulation uses a clustering technique to accelerate RNA folding stochastic simulations.

Another heuristic model was proposed by [32]. Iterative Loop Matching (ILM) is a heuristic algorithm based on the Loop Matching (LM) algorithm. ILM is based on the combination of thermodynamic and comparative approach. It works on a single sequence. It takes as an input either an individual sequence or a sequence alignment. Pseudoknots are considered an interaction between two loop regions, so a number of iterations of LM algorithm are performed. The base pairs corresponding to secondary structure formed in a particular iteration are removed from the original sequence for the next iteration. During the iterations, only base pairs with the highest score are considered. A base-pair scoring matrix from sequence or sequence alignment is prepared by combining a covariance score and thermodynamic score. For individual sequences, the highest score calculation ignores covariance score. This heuristic algorithm also does not predict optimal structures. The algorithm is not too complex, works well on small sequences but is resource demanding. A number of models existed for predicting RNA pseudoknots, but it was soon realized that those models should be enhanced to predict occurring H-type pseudoknots. So, another model, HPknotter [63], was proposed combining some of the existing models, namely RNAMotif [64], pknots [22], Nupack [25] and pknotsRG [26]. The advantage of this combined model is its improved efficiency in predicting actual H-type pseudoknots by applying various filtering mechanisms from time to time in the prediction process.

Another researcher proposed the heuristic model hotknots [33], which is an iterative procedure based on free energy minimization to predict an RNA secondary structure including pseudoknots. Hotknots were computed from an input sequence using a simple local alignment algorithm and then modified in order to keep the total length of hotknots from growing too large. A hotknot is defined as a stem structure composed of stacked pairs, bulge loops containing one unpaired base and interior loops with two (opposing) unpaired bases. A tree is made of a selected set of hotknots (good hotknots), each node of which having a set of hotknots is expanded into a secondary structure. In order to output a pseudoknot free secondary structure with minimum free energy, another algorithm, SimFold [65], was used. This algorithm also uses thermodynamic parameters of Turner [49,82], together with those of [25] for pseudoknotted loops to determine energy of structure and to prune a tree of partial structures. The heuristic hotknots model performs poorly in predicting optimal structures with respect to underlying energy model. Further, the running time of this model also needs to be improved. Inspired by the hotknots model, another iterative model based on evolutionary information as well as thermodynamics was proposed [66]. The performance of this model is better than hotknots in terms of sensitivity and specificity. Another heuristic attempt was made by [67]. Their implementation, Pkiss, combines dynamic programming and canonization rules of [26] to reduce search space thereby lowering the space and time complexity. This heuristic model focused on kissing hairpin pseudoknots and is implemented in algebraic dynamic programming. This model was further improved [68]. Shape analysis and probabilities have been added to improve the prediction accuracy. Another heuristic pseudoknot prediction model [69] was proposed in 2007. They named it vsFold5 and it is available at http://www.rna.it-chiba.ac.jp/vsfold5. vsFold5 make use of biological features like polymer solvent interactions and Kuhn length to predict pseudoknots. Kuhn length is a measure of stiffness. The said model has the tendency to fold according to the thermodynamically most-probable folding pathway. It works for sequence lengths up to 450 nucleotides. The model takes as input, RNA sequence, minimum stem length and number of contiguous stems. Default values of most of the parameters are given. Polymer model options are also given. The output secondary structures are of minimum free energy based on Kuhn length, temperature and polymer solvent prominence irrespective of the fact that a pseudoknot is present or not. Pseudoknot structures have been categorized as core pseudoknots and extended pseudoknots. Extended pseudoknots are formed after secondary structures are formed, while core pseudoknots are formed before the formation of secondary structure.

The worst case time complexity of vsFold5 corresponds to the sequence length. In this model, automated estimation of Kuhn length is an important factor. Also, their future work will be to accommodate parameters such as protein interactions, sub-optimal structures, etc. Another iterative model, Cyclofold [70], was proposed in 2010. This model folded the RNA sequence by choosing the helices iteratively on the basis of free energy. It also checked steric feasibility of these chosen helices to improve accuracy. Steric feasibility relates to the energy produced by atomic distance in a molecule. It can predict pseudoknots of many types. Another model, NanoFolder, was proposed in 2011 [71]. This model focuses on multi-strand RNA secondary structures with multiple crossing arcs. Scoring function uses base-pair stacking energy. A penalty score is introduced for base pairs amongst similar strands. The penalty score amongst two different strands is calculated from the training set. The model collects all the helices with base pairs and calculates their score. Nanofolder takes input as an RNA sequence in FASTA format and outputs the most probable structure, only in Dot-bracket notation. It is available at https://matchfold.ncifcrf.gov.

6.4   FORMAL GRAMMAR TECHNIQUE

Formal grammar based pseudoknot prediction models are mathematical models based on formal grammar. Formal grammar is being used to predict the presence of a pseudoknot in an RNA sequence [37]. Different types of grammar like context free grammar, Multiple Context Free Grammar (MCFG) and Stochastic Context Free Grammar (SCFG) have been used to predict pseudoknots. The inclusion relationship between context free language, regular language, tree adjoining language, multiple context free language and many more grammars have been presented by authors of [72]. Most of these grammars are elaborated in Chapter 7.

A rule-based model was proposed by [73]. This model is based on MFold [74] and used Maude for a pseudoknot structure prediction. Maude is related to term rewriting. RNA structures treated as terms and rules are discovered for structure prediction. First, pseudoknot-free structures are predicted and parsed using term rewriting, then score-based motif-motif interaction is done using term rewriting to predict pseudoknotted structures. The algorithm performs well for average-length sequences but can handle interaction between two unpaired regions of RNA sequence only. Another formal grammar model was proposed by [35]. This model is an attempt to improve the performance of [23]. It uses two types of grammars to predict RNA structures including pseudoknot, simple linear tree adjoining grammars and extended simple linear tree adjoining grammars. It uses a sub-model of TAG named TAGRNA for the prediction of pseudoknots. It considers H-type pseudoknots, including the interaction of two pseudoknots or an interaction of a hairpin loop and a pseudoknot. The scoring function used is based on the minimum free energy concept.

Another attempt to model pseudoknots was made by [75] using a specific language for modelling RNA secondary structures. Various motifs of RNA have been termed tokens. Tokens have been used in formal grammar language to identify RNA structures using CFG. Their specification may be helpful in modelling simple structures only. Also, it works for sequence lengths up to 50 only. Contrary to the context free grammar, context sensitive grammar considers the context in which it is situated. Because of this, it is easy to model various RNA structures. Unfortunately, there is no known automatic parser for CSG. This researcher has tried to parse the said CSG by an augmented transition network. They haven't explained how their model handles pseudoknots explicitly. A subclass of TAG named TAGRNA was proposed by [76] to model pseudoknots. Initially, the time complexity of TAGRNA was $O(n^5)$. So, they constructed a Stochastic Regular Grammar (SREG) to filter the candidate structures, thereby reducing the time complexity. Stochastic parameters are calculated using a logarithm of probability values. For testing the efficiency of system, two types of grammars are generated, one with optimized probability parameters and the other with un-optimized probability parameters. Most RNA pseudoknot prediction models are developed in order to reduce the time and space complexity. In contrast, Gfold [36] was developed to define a more suitable class of structures that can be built recursively. Gfold is also based on MCFG. Apart from the benefits of MCFG, Gfold-incorporated topological aspect of RNA structure, i.e. genus. The genus of an RNA structure is defined to be the minimum g such that the disk diagram can be drawn on a surface of genus g with no crossing arcs. Pseudoknot-free structures have a genus zero. For concatenated structures as well as nested structures, the genus is the sum of two substructures. Using this recursive structure formation, Gfold is implemented using dynamic programming. The concept of genus can be used to topologically classify RNA secondary structures including complex pseudoknots and also to predict the number of secondary structures in an RNA sequence.

Another attempt was made by [77] for the prediction of a pseudoknot using SCFG. Sub-structures are first predicted using SCFG, and then combined to predict pseudoknots and other motifs. The algorithm is time efficient but can be applied only to those homologous RNA sequences that match the training sequences used by the algorithm.

6.5   INTEGER PROGRAMMING TECHNIQUE

Integer programming is a technique based on linear programming, which is defined as the optimization of a linear function subject to a set of linear constraints over integer variables. The RNA pseudoknotted structure prediction problem was formulated as an integer programming problem by [40]. The aim was to minimize the value of the objective function that reflects thermodynamic free energy of a folding structure of an input RNA sequence. The inner base pair of a pseudoknot opens before the helices forming that pseudoknot open. The advantage of using integer programming for a pseudoknot prediction is that various integer programming software systems like CPLEX software [78], LINDO API 6.0 [79] are available for solving the integer programming problem, once the prediction problem is modelled by integer programming. This model showed good performance in terms of accuracy only for RNA sequences of small lengths. Another integer programming model to solve the RNA pseudoknot prediction problem was proposed by [41]. Sato's model Ipknot could work for longer sequences, up to 500 nucleotides. They introduced a concept of threshold cut while finding the Maximum Expected Accuracy (MEA) structure using integer programming. This threshold cut was introduced in order to run the algorithm faster. This algorithm could run for both single sequence and comparative sequence analysis. The accuracy of MEA-based methods is significantly higher when using a parameter set of [60] than using parameters of [14]. Another attempt to predict pseudoknots through integer programming was made by [80]. The structure with the highest number of base pairs is the output structure. Their output structure has minimum free energy. It works for a partial class of pseudoknots.

6.6   INVERSE FOLDING TECHNIQUE

Inverse folding of an RNA sequence is a different concept in which the target structure is converted into an RNA sequence. It was proposed by [39]. INV is an extension of traditional inverse folding models to include a canonical pseudoknot structure prediction. These models take as input the target structure and outputs RNA sequence. INV is a local stochastic search for prediction of a non-bipolar structure, three non-crossing RNA structures with at most, two crossings. It obtains an RNA sequence from a given structure by identifying the specific loop decompositions. The total energy of a structure is calculated by the sum of energies of its substructures. All the sequences that compose a structure can be realized by a neural network. Another inverse folding model [81] was proposed recently. This model uses Fatgraph for the said conversion. Fatgraph takes the secondary structure as the backbone or ribbon and cyclic order on the edge's incident on a vertex. This model makes use of context free grammar and the energy model of Turner for finding the target RNA sequence. The complexity of the model is related to the concept of genus as with base pairs their complexity is exponential.

APPENDIX

Exercise
Multiple-Choice Questions

Q1. In an RNA secondary structure, the energy necessary to form individual base pairs is not quite affected by adjacent base pairs.

    a. True
    b. False

Q2. If the base pair is adjacent to loops or bulges, the neighbouring loops and bulges tend to _____ the base-pair formation.

    a. have no change on
    b. decrease the energy of
    c. stabilize
    d. destabilize

Q3. The dot matrix method and the dynamic programming method can be used in detecting self-complementary regions of a sequence.

    a. True
    b. False

Q4. The most widely explored methods for RNA secondary structure prediction are based on a minimum free energy approach and comparative approach, respectively.

    a. True
    b. False

Q5. To date, standard thermodynamic parameters for a pseudoknot do not exist.

    a. True
    b. False

Q6. Regarding an RNA secondary structure prediction, pick the incorrect statement.

    a. A matched row in anRNA matrix is a representation of complementary nucleotides.
    b. Every base is compared to other complimentary bases available, just like the dot matrix analysis.
    c. If part of the columns and rows match in the RNA matrix, it represents failure of a complimentary nucleotide.
    d. Possible base pairs are GC, GU and AU.

Q7. SCFG is the stochastic version of CFG, similar to the probability theory.

    a. True
    b. False

Q8. In the inverse folding technique of an RNA secondary structure, a prediction target structure is converted into an RNA sequence.

    a. True
    b. False

Q9. Time complexity of the dynamic programming method is reasonable and practical.

    a. True
    b. False

Q10. Sensitivity refers to the ability to find as many correct hits as possible.

    a. True
    b. False

Essay-Type Questions

Q1. Discuss the role of pseudoknots in an RNA secondary structure prediction.

Q2. Various pseudoknot prediction models have been proposed by researchers over a period. Categorize these models on the basis of the algorithm used.

Q3. Compare the RNA secondary structure prediction models on the basis of a few parameters like specificity, sensitivity, etc.

Q4. How can we say that heuristic models are not always applicable in all situations?

Q5. How can inverse folding help in predicting an RNA secondary structure?

7 Pseudoknot Grammar

7.1 CHOMSKY HIERARCHY OF GRAMMARS

A grammar can be defined by four tuples (V, T, P, S) where V is the set of non-terminals, T stands for a set of terminals, P stands for a set of production rules and S stands for start symbol. A symbol is a basic building block in a language, e.g. 0 and 1. The number of symbols is predefined in any grammar. Grammar has a corresponding language. A finite non-empty set of symbols is called alphabet, e.g. {0,1}. The finite sequence of symbols is called a string, e.g. 001. Finally, a set of strings is called a language, e.g. 0n1m where n is even and m is odd that may generate 001, 000011, etc.

As the production rules are applied on S, the construction of a derivation tree starts. The end nodes or leaf nodes of a derivation tree are called terminals. All the other nodes of the tree except leaf nodes are called non-terminals. The start symbol S is a special non-terminal that starts the derivation. A typical derivation tree for grammar G is shown in Figure 7.1.

In the 1956, Noam Chomsky classified grammar into four types: Type 0, Type 1, Type 2 and Type 3 grammar. The properties of grammar are such that they are strictly nested, as shown in Figure 7.2. Type 3 is the most restrictive grammar and Type 0 is the least restrictive. Each set of grammar is characterized by a set of rules. Also, each type has its corresponding language i.e. set of strings generated by grammar and associated automata. The types of grammar of the Chomsky hierarchy are shown in Figure 7.3.

Also, $Type_{i+1} \subset Type_i$ where i = 0, 1, 2 i.e. Type 3 grammar is a subset of Type 2, Type 1 and Type 0 grammar and so on. In other words, if a grammar is a regular grammar, it is automatically context free, context sensitive, and unrestricted grammar.

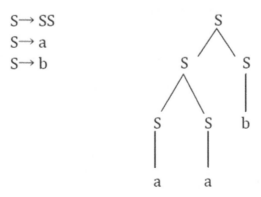

FIGURE 7.1   Derivation Tree of Given Grammar

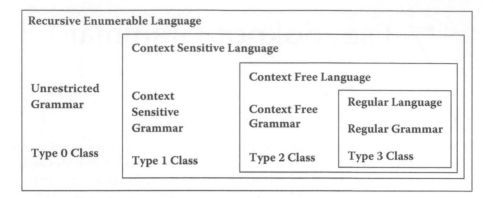

FIGURE 7.2 Chomsky Hierarchy of Grammars

CLASS	GRAMMAR	LANGUAGE	AUTOMATON
Type 3	Regular	Regular	Finite
Type 2	Context Free	Context Free	Pushdown
Type 1	Context Sensitive	Context Sensitive	Linear Bound
Type 0	Unrestricted	Recursive Enumerable	Turing Machine

FIGURE 7.3 Types of Grammar in Chomsky Hierarchy

But the reverse is not true. If grammar is context sensitive, it is not necessary that it is context free.

Type 0 grammar or unrestricted grammar can have productions of the form:

$$A \rightarrow B$$

where A can be terminal as well as non-terminal or a combination of both de-noted by {V∪T}+ and B can be a null, terminal, non-terminal or a combination of both terminals and non-terminals {V∪T}*. In other words, both sides of production rules may contain a combination of terminals and non-terminals but the left-hand side will not include a null or empty symbol.

In Type 1 grammar or context sensitive grammar, the left-hand side and right-hand side of the production rule belongs to {V∪T}+ but there is a further restriction. The length of the left-hand side is less than or equal to the right-hand side i.e.

$$|A| < |B| \text{ or } |A| = |B|$$

This is why these grammars are also called non-contracting grammars.

The most common form of grammar is Type 2 grammar or context free grammar. In this grammar, the right-hand side can be any combination of terminals and non-terminals but the left-hand side contains a single non-terminal i.e. B ∈ {V∪T}* but A ∈ {V}.

A grammar is a regular grammar if it is a left linear grammar or right linear grammar but not both. A left linear grammar is the one whose derivation tree is right skewed. Similarly, a right linear grammar is the one whose derivation tree is left skewed. A middle-linear grammar is not regular. If grammar is regular it is also linear, but the reverse is not true. For example, A → aA | a produces a right skewed tree, and A → Aa | a produces a left skewed tree. Similarly, A → aAb | a produces a tree that grows in the middle.

As restrictions on grammar increases from Type 0 to TYPE 3, their corresponding languages become simple from Type 0 to Type 3. In other words, the more complicated the grammar, the least complicated is its corresponding language.

7.2 CONTEXT FREE GRAMMAR

The class of context free languages generalizes over the class of regular languages, i.e. every regular language is a context free language.

The reverse of this is not true, i.e. every context free language is not necessarily regular. For example, as we will see $\{0^k1^k \mid k \geq 0\}$, is context free but not regular.

Formally, a Context Free Grammar (CFG) is a 4-tuple:

$$|A| < |B| \text{ or } |A| = |B|$$

V − A finite set of variables or non-terminals

T or Σ − A finite set of terminals (V and T do not intersect).

P − A finite set of productions, each of the form A –> α, where A is in V and α is in (V ∪ T)*

Note that α may be ε

S − A starting non-terminal (S is in V)

Example CFG:

$$G = (\{A, \ B, \ C, \ S\}, \{a, \ b, \ c\}, \ P, \ S)$$

P:

   1. S –> ABC
   2. A –> a   A A –> aA | ε
   3. A –> ε
   4. B –> bB   B –> bB | ε
   5. B –> ε
   6. C –> c   C C –> cC | ε
   7. C –> ε

Example Derivations:

$$S = >ABC$$
$$= > BC$$
$$= > C$$
$$= > \varepsilon$$

$$S = >ABC$$
$$= > aABC$$
$$= > aaABC$$
$$= > aaBC$$
$$= > aabBC$$
$$= > aabC$$
$$= > aabcC$$
$$= > aabc$$

Context Free Grammar (CFG) has been in focus for an RNA secondary structure prediction including pseudoknots. A pattern matching technique using stochastic context free grammar was proposed by [82]. Probabilities are assigned to rules of the said CFG model to convert it to a stochastic context free model. Probabilities are calculated by extending the covariance model. Emission and transmission probabilities have been assigned to grammar rules. Multiple sequence alignment is used to find probable motifs of RNA. Their concept divides the secondary structure into various substructures which they call related residues and nonrelated residues. As the simultaneous analysis of structures is not done, they are able to reduce the time complexity to $O(n^2 \log n)$.

Another attempt was made by [77] for the prediction of a pseudoknot using SCFG. Sub-structures are first predicted using SCFG, and then combined to predict pseudoknots and other motifs. The algorithm is time efficient but can be applied only to those homologous RNA sequences that match the training sequences used by the algorithm.

7.3   PARALLEL COMMUNICATING GRAMMAR SYSTEM

In 2003, the Parallel Communicating Grammar System (PCGS) was introduced [83]. It consists of more than one Chomsky grammar, one of which is the master grammar.

This grammar has been used to solve an RNA secondary structure problem. Grammars for various RNA motifs have been introduced here that are controlled by a master PCGS. The RNA motifs are treated as components that are replaced by sub-sequences of RNA in the query string forming a P-structure. The P-structure itself needs to be non-pseudoknotted. This grammar system is then converted into a stochastic version by applying probabilities to grammar rules. These probabilities have been calculated with the help of the CYK algorithm. As the P-structures themselves need to be non-pseudoknotted, it puts a severe restriction on the type of pseudoknots being predicted. Also, parallel grammars need to consider conditional probabilities because of the presence of multiple grammar components that communicate amongst each other. The time complexity of the algorithm is $O(n^6)$ and space complexity is $O(n^4)$.

7.4   CONTEXT SENSITIVE GRAMMAR

A context sensitive grammar is an unrestricted grammar in which all the productions are of the form:

$$\alpha \to \beta, \quad \alpha, \quad \beta \in (V \cup T)^* \text{ and } |\alpha| <= |\beta|$$

where α and β are strings of non-terminals and terminals, respectively.

Context sensitive grammars are more powerful than context free grammars because some languages can be described by CSG but not by context free grammars and CSL are less powerful than unrestricted grammar. That's why context sensitive grammars are sited between context free and unrestricted grammars in the Chomsky hierarchy.

Context sensitive grammar has 4-tuples. G = {N, Σ, P, S}, Where

N = Set of non-terminal symbols

Σ = Set of terminal symbols

S = Start symbol of the production

P = Finite set of productions

All rules in P are of the form $\alpha1$ A $\alpha2 \to \alpha1$ β $\alpha2$ where $\alpha1, \alpha2 \in (N \cup \Sigma)$ and $\beta \in (N \cup \Sigma)+$

The production S $\to \epsilon$ is also allowed if S is the start symbol and it does not appear on the right side of any production.

CSG is a noncontracting grammar. The language generated by the context sensitive grammar is called context sensitive language. If G is a context sensitive grammar, then L(G) = {w | (w $\in \Sigma$ *) $\wedge$ (S $\Rightarrow+_G$ w)}.

The following grammar (G) is context sensitive.

$$S \to aTb|ab$$
$$aT \to aaTb|ac$$
$$L(G) = \{ab\} \cup \{a^n cb^n | n > 0\}$$

This language is also context free.

7.5 PAIR STOCHASTIC TREE ADJOINING GRAMMAR

As we have seen, context free grammar cannot capture all linguistic phenomena and on the other hand the parsing time taken by context sensitive grammar is exponential, so the discussion of tree adjoining grammar becomes crucial here. Tree adjoining grammar is similar to context free grammars but the elementary unit of rewriting is the tree instead of the symbol.

Tree adjoining grammar uses two types of elementary trees called initial trees and auxiliary trees. The rules in a TAG are trees with a special leaf node known as the foot node, which is anchored to a word. Auxiliary trees have the root (top) node and foot node labeled with the same symbol. A derivation starts with an initial tree, combining via either substitution or adjunction. Substitution replaces a frontier node with another tree whose top node has the same label. The root/foot label of the auxiliary tree must match the label of the node at which it adjoins. Adjunction can thus have the effect of inserting an auxiliary tree into the center of another tree (Figure 7.4).

Formally, tree adjoining grammar can be expressed with the following tuples.

G (N, Σ, S, I, A) where N denotes a finite set of no terminals, Σ denotes a finite set of terminals, S denotes the start symbol that represents the whole sentence (S ∈ N). I denote the finite set of initial trees and A denotes the finite set of auxiliary trees. Production rules include substitution and adjunction.

7.5.1 SUBSTITUTION

Substitution is only for initial trees. The process of TAG substitution is as shown in Figure 7.5.

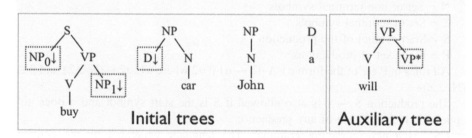

FIGURE 7.4   TAG Initial and Auxiliary Tree

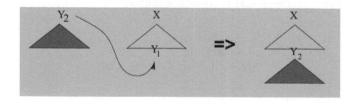

FIGURE 7.5   TAG Substitution

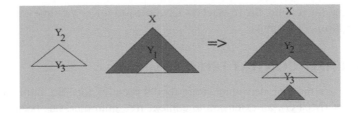

FIGURE 7.6 TAG Adjunction

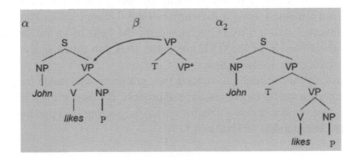

FIGURE 7.7 TAG Adjunction into Initial Tree

7.5.2 ADJUNCTION

Adjunction is for auxiliary trees. The process of TAG adjunction is as shown in Figure 7.6.

An example of the adjunction of T into an initial tree is shown in Figure 7.7.

Tree adjoining grammars are more powerful than context free grammar but less powerful than context sensitive grammar.

A subclass of context sensitive grammars called Pair Stochastic Tree Adjoining Grammar (PSTAGs) has been used to predict pseudoknots [34]. They extended the Pair Hidden Markov models on tree structures [84] to include pseudoknots. They have defined two subclasses of TAG – Simple Linear TAG (SLTAG) and Extended Simple Linear TAG (ESLTAG). These subclasses are realized with initial trees and adjunct trees. RNA substructures can be represented with the help of these trees by applying various operations like adjoin, derivation, etc. PSTAGs work on the structural alignment concept. It takes as input a pair of linear sequences and a TAG representation of a pseudoknot is produced. Using recurrence relation, a dynamic parser produces structural alignment. Based on probability assigned to all alignments of the TAG tree, the most probable alignment is computed. The advantage of this method is that no training of the model is required. The prediction accuracy of this algorithm can be enhanced by using more powerful grammatical methods at the expense of resource cost. The time complexity of the algorithm is $O(KnN^4 + KmN^5)$, where m and n are the numbers of nodes, K is the number of states in PSTAG and N is the sequence length. Tree adjoining grammar has enough computational capacity to model pseudoknots, but it takes too much time for parsing. So, researchers have tried to lower the time complexity by using other forms of grammar.

7.6 MULTIPLE CONTEXT FREE GRAMMAR

Context free grammar alone cannot easily model pseudoknots. So, multiple context free grammar (MCFG), an extension to context free grammar, has been applied to an RNA pseudoknot prediction [85]. MCFG is an extension to CFG to include functions such as bifurcation, base pairing, unpaired based, etc. Each rule in MCFG is accompanied by two types of probabilities, transition probability and emission probability. The most probable tree is found by using the CYK algorithm. CYK algorithm is implemented using a five-dimensional dynamic programming matrix. MCFG is converted into a SMCFG version with the help of an inside-outside algorithm. As their version of the CYK algorithm took too much space, so hashing was used to reduce the space complexity.

Another attempt using multiple context free grammar was made by [38]. Their parsing algorithm is based on comparative approach. The best part of his algorithm is that it does not require training data for parameter setting of a parsing algorithm. His results are comparable to Hxmatch [30].

7.7 PATH CONTROLLED GRAMMAR

Path controlled grammar (PTCG) is a new class of grammar that generates trees restrictively. It was introduced by [86] and further explored by [87] and [88]. Path controlled grammars generate trees on lines of a specific pre-stated path. In other words, the tree generation will be catered by a designated predefined path for a given context free grammar GR.

Path controlled grammar is more powerful than context free grammar but it still preserves the basic properties of context free languages (CFLs). For a given CFG $\bar{G}$, CFL $\acute{L}$ and sequence x,

$x \in \bar{G}$ & $x \in \acute{L}$ iff $\bar{G}$ derives x with a derivation tree T such that there exists a path for T in $\acute{L}$

For designing a path controlled grammar, both CFG and CFL are required. The idea is to impose a restriction on the paths in the derivation tree.

Context free grammars have moderate generating power that is not sufficient to model complex structures like pseudoknots. A class of grammar with better generative power is context sensitive grammar. But the issue with the usage of context sensitive grammar to model pseudoknots are a multifold increase in complexity of the algorithm. So, the idea is to enhance the generative power of CFG without making it too complicated to implement. The solution is PTCG. Without adding context to the grammar, the generative capacity is increased only by controlling the path generation of the derivation tree.

Further, the number of output trees is controlled as only those trees have generated that match the given CFL corresponding to the pseudoknot structure.

APPENDIX

Exercise
Multiple-Choice Questions

Q1. Two features of the tRNA molecule that are associated in converting the triplet codon to an amino acid are

    a. in the T loop and D stem and loop.
    b. in the anticodon loop and D stem loop.
    c. in the anticodon loop and the 3' CCA end.
    d. none of these.

Q2. Which of the following statement is false?

    a. In the derivation tree, the label of each leaf node is terminal.
    b. In the derivation tree, the label of all nodes except leaf nodes is a variable.
    c. In the derivation tree, if the root of a sub-tree is X then it is called – tree.
    d. None of these.

Q3. Regular grammar is

    a. a CFG.
    b. a non-CFG.
    c. an English grammar.
    d. none of these.

Q4. The following grammar

    $G = (N, T, P, S)$
    $N = \{S, A, B\}$
    $T = \{a, b, c\}$
    $P: S \rightarrow aSa$
    $S \rightarrow aAa$
    $A \rightarrow bB$
    $B \rightarrow bB$
    $B \rightarrow c$ is

    a. is type 3.
    b. is type 2 but not type 3.
    c. is type 1 but not type 2.
    d. is type 0 but not type 1.

Q5. The traditional CYK algorithm can be applied directly on path controlled grammar to model pseudoknots.

    a. True
    b. False

Essay-Type Questions

Q1. Explain the Chomsky hierarchy of formal grammars.

Q2. What is TAG? Discuss the two operations associated with it.

Q3. Elaborate with an example of how pseudoknots can be represented using path controlled grammar.

Q4. Discuss a few grammar systems that have generative power more than context free grammar.

Q5. Identify a few models that use formal grammar for predicting pseudoknots.

8 New Areas of Bioinformatics

8.1 DRUG TARGET IDENTIFICATION

Drug discovery starts with initial steps of target identification and passes to the later stages of lead optimization. Multiple sources aid in determining an effective disease target, including academic studies, clinical work and the commercial sector. The target chosen is then used by the pharmaceutical industry or researchers to classify molecules to generate effective drugs. The process involves various early steps, as shown in Figure 8.1.

8.1.1 TARGET IDENTIFICATION AND VALIDATION

Target identification and characterization is an important step. It begins with identifying the function of a possible therapeutic target (gene/protein) and its role in the disease. One such related field is RNA therapeutics. RNA therapeutics drugs are designed to increase or decrease the production of a protein involved in a disease. It is a mechanism to regulate normal gene expression to eradicate several disorders like cancer, Hepatitis C, Huntington Disease, HIV, malaria, etc. The characterization of the molecular pathways discussed by the target is accompanied by recognition of the target. A good target should be efficacious, safe, meet clinical and commercial requirements and be "druggable". The concept of RNAi leads to gene silencing, which means switching off the target gene. It prevents the expression of a certain gene. One of the applications of the gene silencing technique was to cure Huntington disease which is caused by the production of protein called Huntington.

Various approaches for drug target identification are:

1. Data mining using bioinformatics which includes identifying, selecting and prioritizing potential disease targets.
2. Genetic association, which deals with genetic polymorphism and connection with the disease. For example, cystic fibrosis which is a genetic disease that affects the lungs and digestive system. Cystic fibrosis is being treated with the help of LUNAR technology.
3. Expression profile relates to changes in mRNA/protein levels. As gene expression is the process of formation of proteins, wherein DNA acts as a template for mRNA which further gets converted into proteins. So, altering the gene expression process may prevent some diseases from occurring. More precisely, it is altering gene coding at the mRNA level.

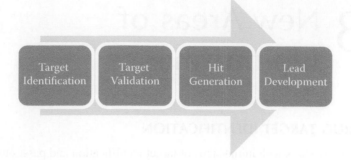

FIGURE 8.1   Steps for Drug Target Discovery

8.1.2   TOOLS FOR TARGET IDENTIFICATION AND VALIDATION

There are various tools and techniques for target identification and validation. They are discussed as follows:

1. Genetics and expression changes: They relate to disease association. Gene expression is the process of formation of proteins, intermediate steps making the conversion from DNA to mRNA.
2. Gene silencing: The process of prevention of gene expression process during translation or transcription process is called gene silencing.
3. LUNAR technology: LUNAR stands for Lipid-enabled and Unlocked Nucleic Acid modified
4. RNA delivery technology: In this technique, a lipid particle is fused into the cell membrane that leads to the biodegradation process and delivers the required RNA.
5. PMO: PMO (phosphorodiamidate morpholino oligomer) is a synthetic RNA that retards the gene expression in order to prevent cells from making a targeted protein. They are molecular structures that are bound to RNA at a specific position.
6. RNA interference: The mechanism of RNA Interference (RNAi) involves treating the disease-causing gene with specific small interfering RNA (siRNA) that targets the specific messenger RNA (mRNA).
7. RNAi therapeutics: Altering gene coding at the mRNA level is called RNAi Therapeutics. RNA therapeutics drugs are designed to increase or decrease the production of a protein involved in a disease.
8. siRNA: Small interfering RNA is also known as silencing RNA or short interfering RNA. siRNA is formed from double-stranded RNA molecular sequence of length 20–25 base pairs and is used.

8.1.3 TARGET VALIDATION

Target validation relates to the molecular target which when treated can lead to therapeutic effect. The validation of the target needs to be accompanied by a multi-validation approach.

One of the approaches include genetic manipulation of target genes (in vitro), including knocking down the gene (shRNA, siRNA, miRNA), knocking out the gene (CRISPR, ZFNs) and knocking in the gene. Another approach could be the use of antibodies that interact with the target with high affinity and block further interactions.

Target identification and validation in the current global pandemic COVID-19 caused by the SARS-CoV-2 virus are important. There is an immediate need for the identification of effective drugs to contain the disastrous virus outbreak. Global efforts are already underway at a war footing to identify the best drug combination to address the disease [89]. Some countries are doing trials on the vaccine. Recently, India has also announced its first vaccine.

Worldwide, research laboratories and drug development organizations work relentlessly to assess compounds that can inhibit the spread of SARS-CoV-2 in humans. To accomplish this, it is important to identify drug targets first and then identify and evaluate compounds and biologics that can effectively engage these targets and inhibit the spread [89].

It is important to understand the biology, mode of transmission and replication cycle of the virus in order to understand the drug targets and appreciate the ongoing efforts aimed at identifying therapies against SARS-CoV-2. This is particularly important since any successful SARS-CoV-2 therapy should ideally target stages in the life cycle of the virus.

8.2 NUTRIGENOMICS

Based on the definition of genetics and genomics by the WHO nutrigenetics can be defined as nutrition and inheritance studies, while nutrigenomics is the study of the mutual interactions between dietary molecules, genes and gene function. The key distinction between genomics and genetics is that genetics scrutinizes the single gene's functioning and structure, while genomics addresses all genes and their interrelationships in order to understand their collective effect on the organism's growth and development.

Nutrigenomics refers in this sense to both nutrigenetics and nutrigenomics. In order to be considered a valued partner in the genetics and genomics arena and to take full advantage of the many new opportunities that have arisen, nutrition researchers, especially those in the nutrigenomics field, need to make a whole-hearted and concerted effort to incorporate bioinformatics expertise in their toolboxes.

A group of European departments is now establishing a European Union–supported network of excellence, the European Nutrigenomics Organization (NuGO), for the integration and development of nutritional genomics. One major idea is that both the science and application of human nutrition are optimally positioned to benefit from the new genomic approaches to biology, using transcriptomics, proteomics and metabolomics. A key objective of the network will be the development, data warehousing and exploitation of nutrition and health-related bioinformatics for the benefit of European nutrition researchers, and for the community as a whole [90].

8.2.1 PROSPECTS AND APPLICATIONS OF NUTRIGENOMICS

The word "nutrigenomics" was developed to determine the relationship between genes and nutrients. There is a lot of optimism that the new "omics" techniques will lead to a more comprehensive understanding of the impact on gene expression and cell biology of bioactive compounds in foods, such as components in fruits, vegetables and milk. New synergisms can be revealed between different bioactive components. The techniques can be used to research the impact of dietary ingredients on gene expression in blood leukocytes and other cells in human meal studies or intervention studies. This can also contribute to the development of new biomarkers to be used for dietary impact assessment. Also, for the assessment of the effectiveness of therapeutic diets in patients with particular diseases, certain approaches may be relevant in clinical nutrition. Detailed research on selected target genes would be required following the hunt for affected genes.

In practice, the application of nutrigenomics uses several "omics" techniques in various disciplines, including genomics, epigenomics, transcriptomics, proteomics and metabolomics, to examine animal cellular and molecular responses to different dietary nutrients in an independent or integrated way, revealing the global effect of nutrients on animal genomes, methylomes/epigenomes, transcriptomes, proteomes and metabolomes, respectively [91].

8.3   TOXICOGENOMICS

Toxicogenomics is the field of science that deals with the collection, interpretation and storage of information about gene and protein activity within a particular cell or tissue of an organism and response to toxic substances. Toxicogenomics combines toxicology with genomics. It is a subdiscipline of pharmacogenomics. Various environmental exposures, age, organics and individual susceptibility are the factors related to toxicogenomics that are responsible for human diseases. It deals with how a genome responds to environmental stressors and toxicants.

Toxicogenomics include studies of the cellular products controlled by the genome (messenger RNAs, proteins, metabolites, etc.). RNA, proteins and intermediary metabolites (Figure 8.2) have been termed "-omic" technologies, based on their ability to characterize all, or most, members of a family of molecules in a single analysis. The powerful methods used for such analysis are gene expression, microRNA network, etc. These advances present exciting opportunities for improved methods of identifying and evaluating potential human and environmental toxicants, and of monitoring the effects of exposures to these toxicants. It has been anticipated that these new technologies will (1) lead to new families of biomarkers that permit characterization and efficient monitoring of cellular perturbations, (2) provide an increased understanding of the influence of genetic variation on

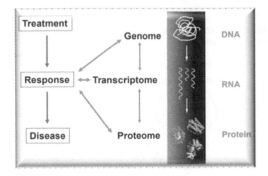

FIGURE 8.2 Toxicogenomics

toxicological outcomes and (3) allow the definition of environmental causes of genetic alterations and their relationship to human disease [92].

The top three phthalate toxicity categories are found to be cardiotoxicity, hepatotoxicity and nephrotoxicity, and the top twenty diseases include cardiovascular, liver, urologic, endocrine and genital diseases.

8.4 BIOTERRORISM

The advances in RNA and DNA have led to their use in areas like identification of victims, paternity test, medical diagnostics, forensic archaeology, DNA fingerprinting, RNA therapeutics, etc. But with these advents, there comes a danger of producing superbugs either accidentally as bioterrorism agents or explicitly for the use of biological weapons.

Biological weapons are the combination of biological agents such as viruses, bacteria and their derivatives like toxins, and the means of keeping the agents alive and virulent, transporting them to where they will be dispersed and a dissemination mechanism.

Bioterrorism is terrorism involving the intentional release or dissemination of biological agents. These agents are bacteria, viruses, insects, fungi or toxins, and may be in a naturally occurring or a human-modified form, in much the same way in biological warfare.

It is a type of warfare that makes use of biological agents or biological weapons to harm the enemy. These weapons use chemicals from either animals or plants or microorganisms. Apart from this, deadly infectious agents may also be used to turn these weapons into bioterrorism agents. These infectious agents may include viruses, protozoa or fungi. Bioterrorism agents may also be modified versions of normally occurring pathogens.

There are three categories of bioterrorism agents. All of the three categories have the potential to be weaponized and they can cause substantial damage to human life.

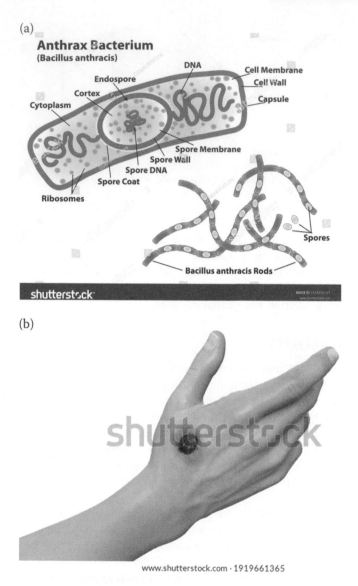

FIGURE 8.3 (a) Category A *Bacillus anthracis*; (b) Category A *Bacillus anthracis*
https://www.shutterstock.com/image-vector/anthrax-bacteria-bacillus-anthracis-endospore-labeled-1555428107
https://www.shutterstock.com/image-illustration/cutaneous-anthrax-most-common-form-infection-1919661365

8.4.1 CATEGORY A BIOTERRORISM AGENTS

This category of bioterrorism agents is the most dangerous. One of the agents that falls in this category is *Bacillus anthracis,* as shown in Figure 8.3(a) and 8.3(b).

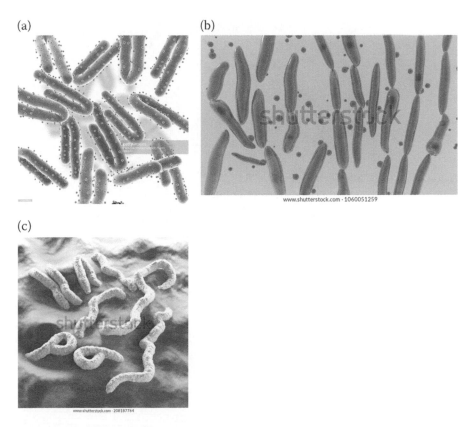

FIGURE 8.4 (a) Marburg Virus; (b) *Clostridium botulinum*; (c) Marburg Virus
https://www.gettyimages.in/detail/illustration/marburg-virus-particles-illustration-royalty-free-illustration/758310809?adppopup=true
https://www.shutterstock.com/image-illustration/botulism-clostridium-botulinum-bacterium-3d-illustration-1060051259
https://image.shutterstock.com/image-illustration/marburg-virus-600w-208187764.jpg

Other agents that fall in category A are *Clostridium botulinum, Yersinia pestis, Francisella tularesnsis,* plague, Ebola, Marburg virus, Machupo virus, variola major, Lassa virus, etc. Some of them are shown in Figure 8.4(a) and 8.4(b), which is taken from "https://www.cbsnews.com/pictures/ore-man-survives-black-death-plague-graphic-images/", and 8.4(c). These agents can result in symptoms like conjunctivitis, headache, deafness, fever, nausea, vomiting or flulike illness. Marburg virus is a single-stranded RNA virus that is linear and non-segmented and was spread through an animal. The Machupo virus caused massive oral bleeding.

8.4.2 Category B Bioterrorism Agents

This category is more dangerous than category C but less dangerous than category A. The death rate due to category B virus is less. They result in moderate illness.

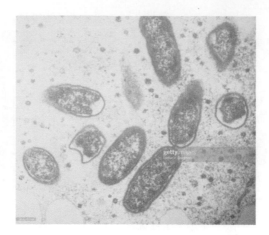

FIGURE 8.5 Typhus Disease – Category B
https://www.gettyimages.in/detail/photo/spotted-fever-bacteria-royalty-free-image/583
679158?adppopup=true

The examples are Glander, Meliodosis, Ricin Q. fever, typhus (Figure 8.5), Staphylococcal, Brucellosis, viral encephalitidies, *Clostridium perfringes* (Figure 8.6), *Ricinus communis,* etc. The symptoms caused by these agents are nasal discharge, abscess formation, acute pneumonia, chronic TB, heart failure, food poisoning, etc.

8.4.3 Category C Bioterrorism Agents

Category C of bioterrorism agents are the ones easily available, easily produced and spread. They include pathogens that could be engineered for mass spread. These agents have the potential to cause a high fatality rate and major death impact.

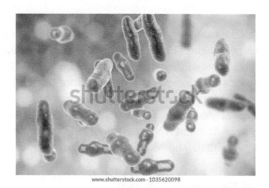

FIGURE 8.6 *Clostridium perfringens*
https://www.shutterstock.com/image-illustration/clostridium-perfringens-bacteria-anaerobic-
sporeproducing-causative-1035620098

www.shutterstock.com · 1684896415

FIGURE 8.7   Hanta Virus
https://www.shutterstock.com/image-photo/text-hantavirus-man-removes-protective-mask-1
684896415

Examples are Nipah virus, Hantavirus (Figure 8.7), Tuberculosis, SARS, Yellow fever, Influenza, Chikungunya, etc. Symptoms include fever, arthralgia, etc.

8.4.4   WHY ARE BIOLOGICAL WEAPONS OR BIOTERRORISM AGENTS SO DISRUPTIVE?

Biological weapons or bioterrorism agents are termed as very disruptive because of the following reasons:

- They can easily cause and transmit the infection.
- Their life expectancy is high.
- Capacity spreading is strong.
- If warehoused, they will be stable.
- They are resilient to environmental conditions.
- They are difficult to trace.
- They are unseen and microscopic.

8.4.5   EFFECTS OF BIOLOGICAL ATTACK

The consequences of biological attacks are very dangerous. According to the World Health Organization:

- Release of 50 kg of Anthrax spores in a 40 km² area will cause 1 lac death in individuals.
- Release of 50 kg of plague spores in a 20 km² area will cause 36,000 people to die.

APPENDIX

Exercise
Multiple-Choice Questions

Q1. Which of the following statements best describes the role of proteins as therapeutic targets?

   a. Very few drugs exert their effects by interacting with proteins.
   b. Drugs targeting enzymes usually activate their target protein.
   c. Drugs often work by enhancing the binding of an enzyme's substrate.
   d. Drugs targeting proteins are often very specific and can be less likely to produce side effects.

Q2. The identification of drugs through the genomic study is called_____.

   a. genomics
   b. pharmacogenomics
   c. pharmacogenetics
   d. cheminformatics

Q3. Proteomics refers to the study of _____.

   a. a set of proteins in a specific region of the cell
   b. biomolecules
   c. a set of proteins
   d. the entire set of expressed proteins in the cell

Q4. Which of the following are not the application of bioinformatics?

   a. drug designing
   b. data storage and management
   c. understand the relationships between organisms
   d. None of these.

Q5. The laboratory work using computers and computer-generated models generally offline is referred to as _____.

   a. insilico
   b. the wet lab
   c. the dry lab
   d. All of these.

Q6. Which statement about the process of drug discovery is true?

   a. It only encompasses the non-clinical laboratory and animal testing.

b. It is the process that ascertains the effectiveness and safety of potential drug candidates.
c. It is the process by which therapeutic compounds are formulated into medicines.
d. It ensures there are no side-effects associated with the potential drug candidates.

Q7. Which statement best describes the distinction between the purposes of the two RNAi pathways within the eukaryotic cell?

a. miRNA-mediated gene silencing represents a back-up pathway should siRNA-mediated gene silencing be unsuccessful at silencing the target dsRNA.
b. siRNA- and miRNA-mediated gene silencing pathways must both be active to successfully silence the target dsRNA.
c. miRNA- and siRNA-mediated gene silencing represent evolutionary-independent pathways that confer identical silencing mechanisms on the target dsRNA.
d. siRNA-mediated gene silencing represents a cell defense mechanism against exogenous dsRNA; miRNA-mediated gene silencing is an integral gene expression regulation process.

Q8. The studies that led to the discovery of post-transcriptional gene silencing phenomena were performed on _____.

a. a petunia
b. a sunflower
c. maize
d. wheat

Q9. RNA interference is evolved as a genetic immune system.

a. True
b. False

Q10. The first RNAi therapeutics was aimed at treating _____.

a. muscular dystrophy
b. macular degeneration
c. multiple myeloma
d. neurodegeneration

Essay-Type Questions

Q1. Correlate RNA interference with human diseases.

Q2. RNAs are highly conserved. Do you agree with this statement? Give reasons to support your answer.

Q3. What is the role of gene expression in RNAi therapeutics?

Q4. How can bioterrorism damage the human population?

Q5. Discuss toxicogenomics as a threat to human life.

Glossary

Bioinformatics: Bioinformatics is conceptualizing biology in terms of macro-molecules (in the sense of physical-chemistry) and then applying "informatics" techniques to understand and organize the information associated with these molecules, on a large scale.

Coaxial Stacking of Stems: Long duplexes of RNA stack upon each other to form a long chain.

Computational Biology: Computational biology involves writing algorithms and code for developing those tools that are to be used by bioinformaticists.

Epigenomics: Studies of hereditary marks in chromatin (histones, DNA)

Gene Expression: Foundation of all the chemical reactions in our body

Gene Silencing: The process of prevention of the gene expression process during translation or transcription process is called gene silencing.

Genetic variation: Studies of genome variations

Genomics: Studies of genomes and functional and regulatory elements

Helix: A helix is formed by stacking base pairs.

H-type Pseudoknot: The most common type of pseudoknot is the H-type pseudoknot.

Metabolomics: Studies of metabolites in cells, tissues, and body fluids

Non-coding RNA (ncRNA): These are functional molecules of RNA that do not encode into protein.

Nutrigenetics: Nutrigenetics can be defined as nutrition and inheritance studies.

Phylogenetics: Phylogenetics is a science of studying biological relationships between organisms that have emerged from a common ancestor over a period of time.

Proteomics: Studies of proteins, including their structure

Pseudoknot: Important motif in RNA secondary structure

RNA Interference: The mechanism of RNA Interference (RNAi) involves treating the disease-causing gene with a specific small interfering RNA (siRNA) that targets the specific messenger RNA (mRNA).

Sequence Alignment: Sequence alignment is the process of placing two or more sequences one over the other such that regions of maximum similarity can be located.

Snurps: Conserved RNA-protein complexes

Species: Species is defined as a set of similar organisms.

Systems Biology: Holistic analysis of the cellular biochemical interaction networks

Toxicogenomics: Toxicogenomics is the field of science that deals with the collection, interpretation and storage of information about gene and protein activities within a particular cell or tissue of an organism and response to toxic substances.

Transcriptomics: Studies of transcripts, including non-coding RNA and microRNA

IMPORTANT DATABASES

EMBL-Bank: www.ebi.ac.uk/embl/
Ensembl: www.ensembl.org/
Ensembl-Genomes: www.ensemblgenomes.org/
GenBank: www.ncbi.nlm.nih.gov/GenBank
Genomes Project: www.1000genomes.org/
KEGG Compound: www.genome.jp/kegg /compound
PseudoBase: http://www.ekevanbatenburg.nl/PKBASE/PKB.HTML
PseudoBase++: https://www.ncbi.nlm.nih.gov/pmc/articles/PMC2686561/
RNACentral: https://rnacentral.org/
tRNAdb: trnadb.bioinf.uni-leipzig.de/1000

References

[1] S. D. Mooney, J. D. Tenenbaum, and R. B. Altman, "Bioinformatics", In: Shortliffe E., Cimino J. (eds), *Biomedical Informatics*. Springer, London, pp. 695–719, 2013. https://doi.org/10.1007/978-1-4471-4474-8_24

[2] N. M. Luscombe, D. Greenbaum, and M. Gerstein, "What Is Bioinformatics? A Proposed Definition and Overview of the Field", *Methods of Information in Medicine*, vol. 40, no. 4, pp. 346–358, 2001. PMID: 11552348.

[3] D. A. Benson, I. Karsch-Mizrachi, D. J. Lipman et al., "Genbank", *Nucleic Acids Research*, vol. 28, no. 1, pp. 15–28, 2000.

[4] T. Hubbard, D. Barker, E. Birney et al., "The Ensembl Genome Database Project", *Nucleic Acids Research*, vol. 30, no. 1, pp. 38–41, 2002.

[5] A. Wilke, T. Harrison, J. Wilkening et al., "The M5NR: A Novel Non-redundant Database Containing Protein Sequences and Annotations from Multiple Sources and Associated Tools", *BMC Bioinformatics*, vol. 13, pp. 141, 2012.

[6] A. Bairoch and R. Apweiler, "The SWISS-PROT Protein Sequence Database and its Supplement Trembl in 2000", *Nucleic Acids Research*, vol. 28, no. 1, pp. 45–48, 2000.

[7] A. Hamosh, A. F. Scott, J. S. Amberger et al., "Online Mendelian Inheritance in Man (OMIM), a Knowledgebase of Human Genes and Genetic Disorders", *Nucleic Acids Research*, vol. 33 (Database Issue), pp. D514–D517, 2005.

[8] H. M. Berman, J. Westbrook, Z. Feng et al., "The Protein Data Bank", *Nucleic Acids Research*, vol. 28, no. 1, pp. 235–242, 2000.

[9] M. Kanehisa and S. Goto. "KEGG: Kyoto Encyclopedia of Genes and Genomes", *Nucleic Acids Research*, vol. 28, no. 1, pp. 27–30, 2000.

[10] K. Rietveld, R. V. Poelgeest, C. W. Pleij et al., "The tRNA-like Structure at the 3′ Terminus of Turnip Yellow Mosaic Virus RNA – Differences and Similarities with Canonical tRNA", *Nucleic Acids Research*, vol. 10, pp. 1929–1946, 1982.

[11] M. Taufer, A. Licon, R. Araiza et al., "Pseudobase Plus Plus: An Extension to Pseudobase for Easy Searching Formatting and Visualization of Pseudoknots", *Nucleic Acids Research*, vol. 37, pp. D127–D135, 2009.

[12] M. Zuker and P. Stiegler, "Optimal Computer Folding of Large RNA Sequences Using Thermodynamics and Auxiliary Information", *Nucleic Acids Research*, vol. 9, pp. 133–148, 1981.

[13] I. L. Hofacker, W. Fontana, P. F. Stadler et al., "Fast Folding and Comparison of RNA Secondary Structures", *Monatsh Chemistry*, vol. 125, no. 2, pp. 167–188, 1994.

[14] D. H. Mathews, J. Sabina, M. Zuker et al., "Expanded Sequence Dependence of Thermodynamic Parameters Provides Improved Prediction of RNA Secondary Structure", *Journal of Molecular Biology*, vol. 288, pp. 911–940, 1999.

[15] I. Hofacker, M. Fekete and P. Stadler, "Secondary Structure Prediction for Aligned RNA Sequences", *Journal of Molecular Biology*, vol. 319, pp. 1059–1066, 2002.

[16] Y. E. Ding, C. Y. Chan and C. E. Lawrence, "RNA Secondary Structure Prediction by Centroids in a Boltzmann Weighted Ensemble", *RNA*, vol. 11, no. 8, pp. 1157–1166, 2005. doi: 10.1261/rna.2500605

[17] R. Nussinov and A. B. Jacobson, "Fast Algorithm for Predicting the Secondary Structure of Single-Stranded RNA", *Proceedings of the National Academy of Sciences of the USA*, vol. 77, no. 11, pp. 6309–6313, 1980.

[18] A. E. Walter, D. H. Turner, J. Kim et al., "Coaxial Stacking of Helixes Enhances Binding of Oligoribonucleotides and Improves Predictions of RNA Folding", *Proceedings of National Academy of Sciences of the USA*, vol. 91, pp. 9218–9222, 1994.

[19] B. Knudsen and J. Hein. "Pfold: RNA Secondary Structure Prediction Using Stochastic Context-Free Grammars", *Nucleic Acids Research*, vol. 31, no. 13, pp. 3423–3428, 2003.

[20] F. H. D. V. Batenburg, A. Gultyaev, C. W. A. Pleij et al., "Pseudobase: A Database with RNA Pseudoknots", *Nucleic Acids Research*, vol. 28, pp. 201–204, 2000.

[21] G. J. Sam, B. Alex, M. Mhairi et al., "RFAM: An RNA Family Database", *Nucleic Acids Research*, vol. 31, no. 1, pp. 439–441, 2003.

[22] E. Rivas and S. R. Eddy, "A Dynamic Programming Algorithm for RNA Structure Prediction including Pseudoknots", *Journal of Molecular Biology*, vol. 285, no. 5, pp. 2053–2068, 1999.

[23] Y. Uemura, A. Hasegawa, S. Kobayashi et al., "Tree Adjoining Grammars for RNA Structure Prediction", *Theoretical Computer Science*, vol. 210, pp. 277–303, 1999.

[24] T. Akutsu, "Dynamic Programming Algorithms for RNA Secondary Structure Prediction with Pseudoknots", *Discrete Applied Mathematics*, vol. 104, pp. 45–62, 2000.

[25] R. M. Dirks and N. A. Pierce, "A Partition Function Algorithm for Nucleic Acid Secondary Structure including Pseudoknots," *Journal of Computational Chemistry*, vol. 24, pp. 1664–1677, 2003.

[26] J. Reeder and R. Giegerich, "Design Implementation and Evaluation of A Practical Pseudoknot Folding Algorithm Based on Thermodynamics", *BMC Bioinformatics*, vol. 5, pp. 104–115, 2004.

[27] M. Mohl, S. Will and R. Backofen, "Fixed Parameter Tractable Alignment of RNA Structures Including Arbitrary Pseudoknots", *Combinatorial Pattern Matching*, vol. 5029, pp. 69–81, 2008.

[28] X. Huang and H. Ali, "High Sensitivity RNA Pseudoknot Prediction", *Nucleic Acids Research*, vol. 35, no. 2, pp. 656–663, 2007.

[29] M. Mohl, R. Salari, S. Will et al., "Sparsification of RNA Structure Prediction Including Pseudoknots", *Algorithms for Molecular Biology*, vol. 5, no. 1, p. 39, 2010.

[30] C. Witwer, I. L. Hofacker and P. F. Stadler, "Prediction of Consensus RNA Secondary Structures Including Pseudoknots", *Transactions on Computational Biology and Bioinformatics*, vol. 1, no. 2, pp. 66–77, 2004.

[31] S. Bellaousov and D. H. Mathews, "Probknot: Fast Prediction of RNA Secondary Structure Including Pseudoknots", *RNA*, vol. 16, no. 10, pp. 1870–1880, 2010.

[32] J. Ruan, G. Stormo and W. Zhang, "An Iterated Loop Matching Approach to the Prediction of RNA Secondary Structures with Pseudoknots", *Bioinformatics*, vol. 20, no. 1, pp. 58–66, 2004.

[33] J. Ren, B. Rastegari, A. Condon et al., "Hotknots: Heuristic Prediction of RNA Secondary Structures Including Pseudoknots", *RNA*, vol. 11, pp. 1494–1504, 2005.

[34] H. Matsui, K. Sato and Y. Sakakibara, "Pair Stochastic Tree Adjoining Grammars for Aligning and Predicting Pseudoknot RNA Structures", *Bioinformatics*, vol. 21, no. 11, pp. 2611–2617, 2004.

[35] S. Rajasekaran, S. A. Seesi and R. A. Ammar, "Improved Algorithms for Parsing ESLTAGS: A Grammatical Model Suitable for RNA Pseudoknots", *Computational Biology and Bioinformatics*, vol. 7, no. 4, pp. 619–627, 2010.

[36] C. M. Reidys, F. W. D. Huang, J. E. Andersen et al., "Topology and Prediction of RNA Pseudoknots", *Bioinformatics*, vol. 27, no. 8, pp. 1076–1085, 2011.

[37] A. Jiwan and S. Singh. "Soft Computing Based Model for Identification of Pseudoknots in RNA Sequence Using Learning Grammar", *International Journal of Computer Applications*, vol. 54, no. 9, pp. 1–7, 2012.

[38] H. Seki, N. Mizoguchi and Y. Kato, "A Comparative Approach to RNA Pseudoknotted Structure Prediction Based on Multiple Context-Free Grammar", *Proceedings of International Conference AWFS*, Shanghai, China, vol. 13–14, 2011.

[39] J. Z. M. Gao, L. Y. M. Li and C. M. Reidys, "Inverse Folding of RNA Pseudoknot Structures", *Algorithms for Molecular Biology*, vol. 5, pp. 27, 2010.

[40] U. Poolsap, Y. Kato and T. Akutsu, "Prediction of RNA Secondary Structure with Pseudoknots using Integer Programming", *BMC Bioinformatics*, vol. 10, no. 1, pp. S38, 2009. doi: 10.1186/1471-2105-10-S1-S38

[41] K. Sato, Y. Kato, M. Hamada et al., "Ipknot: Fast and Accurate Prediction of RNA Secondary Structures with Pseudoknots Using Integer Programming", *Bioinformatics*, vol. 27, pp. I85–I93, 2011.

[42] R. Achawanantakun and Y. Sun, "Shape and Secondary Structure Prediction of ncRNAs Including Pseudoknots Based on Linear SVM", *BMC Bioinformatics*, vol. 14, Suppl 9, pp. S1, 2013. http://www.biomedcentral.com/1471-2105/14/s2/s1. doi: 10.1186/1471-2105-14-S2-S1

[43] R. Christian, *Combinatorial Computational Biology of RNA: Pseudoknots and Neutral Networks,*. *Springer-Verlag New York*, 2011. ISBN 978-0-387-76731-4 2011. doi: 10.1007/978-0-387-76731-4

[44] D. Lee and K. Han, "A Genetic Algorithm for Predicting RNA Pseudoknot Structures", *Lecture Notes in Computer Science*, vol. 2659, pp. 130–139, 2003.

[45] A. Jiwan and S. Singh. "A Review on RNA Pseudoknot Structure Prediction Techniques", *Proceedings of International Conference on Computing, Electronics and Electrical Technologies (ICCEET)*, pp. 975–978, 2012. doi: 10.1109/icceet.B2012.6203854

[46] S. S. Ray and S. K. Pal, "RNA Secondary Structure Prediction Using Soft Computing", *IEEE/ACM Transactions on Computational Biology and Bioinformatics*, vol. 10, pp. 2–17, 2013.

[47] Y. Uemura and A. Hasegawa, "Grammatically Modeling and Predicting RNA Secondary Structures", *Genome Informatics Workshop*. Universal Academy Press Tokyo, 67–76, 1995.

[48] K. Tong, K. Cheung, K. Lee et al., "Modified Free Energy Model to improve RNA Secondary Structure Prediction with Pseudoknots", *Proceedings of 13th IEEE International Conference on Bioinformatics and BioEngineering*, 2013, doi: 10.1109/bibe.2013.6701532

[49] K. K. Tong, K. Y. Cheung, K. H. Lee et al., "GAknot: RNA Secondary Structures Prediction with Pseudoknots Using Genetic Algorithm", *Proceedings of Computational Intelligence in Bioinformatics and Computational Biology*, pp. 136–142, 2013. doi: 10.1109/CIBCB.2013.6595399

[50] T. Akutsu, "Approximation and Exact Algorithms for RNA Secondary Structure Prediction and Recognition of Stochastic Context-free Languages", *Journal of Combinatorial Optimization*, vol. 3, no. 2–3, pp. 321–336, 1999.

[51] H. Jabbari, A. Condon, A. Pop et al., "HFold: RNA Pseudoknotted Secondary Structure Prediction Using Hierarchical Folding", *Algorithms in Bioinformatics*, vol. 4645, pp. 323–334, 2007.

[52] H. L. Chen, A. Condon and H. Jabbari "An O(n5) Algorithm for MFE Prediction of Kissing Hairpins and 4-chains in Nucleic Acids", *Journal of Computational Biology*, vol. 16, no. 6, pp. 803–815, 2009.

[53] H. Jabbari and A. Codon, "A Fast and Robust Iterative Algorithm for Prediction of RNA Pseudoknotted Secondary Structures", *BMC Bioinformatics*, vol. 15, pp. 147, 2014.

[54] T. Jiang, G. Lin, B. Ma and K. Zhang. "A General Edit Distance between RNA Structures" *„Journal of Computational Biology*, vol. 9, no. 2, pp. 371–388, 2002.

[55] T. Wong, Y. S. Chiu, T. W. Lam et al., "A Memory Efficient Algorithm for Structural Alignment of RNAs with Embedded Simple Pseudoknots", *Journal of Computational Biology and Bioinformatics*, vol. 9, no. 1, pp. 161–168, 2012.

[56] B. Dost, B. Han, S. Zhang et al., "Structural Alignment of Pseudoknotted RNA", Research in *Computational Biology*, vol. 3909, pp. 143–158, 2006.

[57] C. Ma, T. K. F. Wong, T. W. Lam et al., "An Efficient Alignment Algorithm for Searching Simple Pseudoknots over Long Genomic Sequence", *IEEE/ACM Transactions on Computational Biology and Bioinformatics*, vol. 9, pp. 1629–1638, 2013.

[58] J. Sperschneider, A. Datta and M. J. Wise, "Predicting Pseudoknotted Structures Across Two RNA Sequences", *Bioinformatics*, vol. 28, no. 23, pp. 3058–3065, 2012.

[59] J. Sperschneider and A. Datta, "Dotknot: Pseudoknot Prediction Using the Probability Dot Plot Under a Refined Energy Model", *Nucleic Acids Research*, vol. 38, no. 7, p. E103, 2010. doi: 10.1093/nar/gkq021.epub2010jan31

[60] M. S. Andronescu, C. Pop and A. E. Condon, "Improved Free Energy Parameters for RNA Pseudoknotted Secondary Structure Prediction", *RNA*, vol. 16, pp. 26–42, 2010.

[61] H. Jabbari, I. Wark and C. Montemagno, "RNA Secondary Structure Prediction with Pseudoknots: Contribution of Algorithm versus Energy Model", *PloS One*, vol. 13, no. 4, p. e0194583, 2018.

[62] A. Xayaphoummine, T. Bucher, F. Thalmann et al., "Prediction and Statistics of Pseudoknots in RNA Structures Using Exactly Clustered Stochastic Simulations", *PNAS*, vol. 100, no. 26, pp. 15310–15315, 2003.

[63] C. H. Huang, C. L. Lu and H. T. Chiu, "A Heuristic Approach for Detecting RNA H-Type Pseudoknots", *Bioinformatics*, vol. 21, no. 17, pp. 3501–3508, 2005.

[64] T. J. Macke, D. J. Ecker, R. R. Gutell et al., "RNAmotif: An RNA Secondary Structure Definition and Search Algorithm", *Nucleic Acids Research*, vol. 29, no. 22, pp. 4724–4735, 2001.

[65] M. Andronescu, Z. C. Zhang and A. Condon, "Secondary Structure Prediction of Interacting RNA Molecules", *Journal of Molecular Biology*, vol. 345, pp. 987–1001, 2005.

[66] K. C. Wiese and A. Hendriks, "RNA Pseudoknot Prediction via an Evolutionary Algorithm", *International Conference on Evolutionary Computation*, CEC, Trondheim, Norway, pp. 270–276, 2009.

[67] C. Theis, S. Janssen and R. Giegerich, "Prediction of RNA Secondary Structure including Kissing Hairpin Motifs", *Proceedings of the 10th International Conference on Algorithms in Bioinformatics*, WABI'10, Springer-Verlag Berlin Heidelberg, 52–64, 2010.

[68] S. Janssen and R. Giegerich, "The RNA Shapes Studio", *Bioinformatics*, vol. 31, no. 3, pp. 423–425, 2015.

[69] W. K. Dawson, K. Fujiwara and G. Kawai, "Prediction of RNA Pseudoknots using Heuristic Modeling with Mapping and Sequential Folding", *PLOS One*, vol. 2, no. 9, pp. E905 (1–7), 2007.

[70] E. Bindewald, T. Kluth and B. A. Shapiro, "CyloFold: Secondary Structure Prediction Including Pseudoknots", *Nucleic Acids Research*, vol. 38, pp. W368–W372, 2010.

[71] E. Bindewald, K. Afonin, L. Jaeger et al., "Shapiro. Multi-Strand RNA Secondary Structure Prediction and Nanostructure Design Including Pseudoknots", *ACS Nano*, vol. 5, no. 12, pp. 9542–9551, 2011.

[72] Y. Kato, H. Seki and T. Kasami. "A Comparative Study on Formal Grammars for Pseudoknots", *Genome Informatics*, vol. 14, pp. 470–471, 2003.

[73] X. Z. Fu, H. Wang, W. Harrison et al., "A Rule-Based Approach for RNA Pseudoknot Prediction", *International Journal of Data Mining and Bioinformatics*, vol. 2, no. 1, pp. 78–93, 2008.

[74] M. Zuker, "Mfold: Web Server for Nucleic Acid Folding and Hybridization Prediction", *Nucleic Acids Research*, vol. 31, no. 13, pp. 3406–3415, 2003.

[75] K. Y. Sung, "A Software Specification Language for RNA Pseudoknots", *Software Engineering Research and Practice*, CSREA Press, pp. 684–687, 2006.

[76] K. Ogasawara and S. Kobayashi, "Stochastic Regular Approximation of Tree Grammars and its Application to Faster ncRNA Family Annotation", *Information and Media Technologies*, vol. 3, no. 1, pp. 26–36, 2008.

[77] C. Yuan and Y. Sun, "Efficient Known ncRNA Search Including Pseudoknots", *BMC Bioinformatics*, vol. 14, Suppl 2, pp. S25, 2013. http://www.biomedcentral.com/14 71-2105/14/s2/s25

[78] R. Lima, "Presentation on IBM ILOG CPLEX", 2010. http://egon.cheme.cmu.edu/ ewocp/docs/ rlima_cplex_ewo_dec2010.pdf http://www.ilog.com/products/cplex/

[79] LINDO. Documentation of LINDO API 6.0. Lindo System Inc Chicago Illinois, 2009. http://web.ist.utl.pt/~ist11038/compute/or/lindo/lindoapi_part.pdf

[80] G. H. Shirdel and N. Kahkeshani, "An Integer Linear Programming Problem for RNA Structures", *Applied Mathematical Sciences*, vol. 6, no. 54, pp. 2695–2702, 2012.

[81] F. Huang, C. Reidys and R. Rezazadegan, "Fatgraph Models of RNA Structure", *Molecular Based Mathematical Biology*, vol. 5, pp. 1–20, 2017.

[82] R. Garcia, "Prediction of RNA Pseudoknotted Secondary Structure Using Stochastic Context Free Grammars (SCFG)" , *CLEI Electronic Journal*, vol. 9, no. 2, pp. 1–12, 2006.

[83] L. Cai, R. L. Malmberg and Y. Wu, "Stochastic Modeling of RNA Pseudoknotted Structures: A Grammatical Approach", *Bioinformatics*, vol. 19, pp. I66–I73, 2003. doi:10.1093/bioinformatics/btg1007

[84] Y. Sakakibara, "Pair Hidden Markov Models on Tree Structures", *Bioinformatics*, vol. 19, no. 1, pp. I232–I240, 2003.

[85] Y. Kato, H. Seki and T. Kasami. "Stochastic Multiple Context-Free Grammar for RNA Pseudoknot Modeling", *Proceedings of the 8th International Workshop on Tree Adjoining Grammar and Related Formalisms, Association for Computational Linguistics*, Sydney, pp. 57–64, 2006.

[86] S. Marcus, C. Martín-Vide, V. Mitrana et al., "A New-Old Class of Linguistically Motivated Regulated Grammars", *Language and Computers*, vol. 37, no. 1, pp. 111–125, 2001.

[87] M. Cermak, J. Koutny and A. Meduna, "Parsing Based on n-Path Tree-Controlled Grammars", *Theoretical and Applied Informatics*, vol. 23, no. 3-4, pp. 213–238, 2011. doi: 10.2478/v10179-011-0015-7

[88] C. R. Gramatiky, "Path Controlled Grammars", Masters Dissertation, Department of Information Systems: BRNO University of Technology, 2015.

[89] A. Saxena, "Drug Targets for COVID-19 Therapeutics: OngoinGlobal Efforts", *Journal of Biosciences*, vol. 45, 87, 2020. https://doi.org/10.1007/s12038-020-00067-w

[90] B. Åkesson, G. Önning, H. Lindmark Månsson and Å. Nilsson, "Nutrigenomics: New Tools forNutritional Science", *Scandinavian Journal of Nutrition*, vol. 48, no. 2, pp. 95–97, 2004. doi: 10.1080/11026480410027013

[91] M. S. Hasan, J. M. Feugang, S. F. Liao, "A Nutrigenomics Approach Using RNA Sequencing Technology to Study Nutrient–Gene Interactions in Agricultural Animals", *Current Developments in Nutrition*, vol. 3, no. 8, 2019. https://doi.org/10.1093/cdn/ nzz082

[92] M. J. Aardema and J. T. MacGregor, "Toxicology and Genetic Toxicology in the New Era of "Toxicogenomics": Impact of "-omics" Technologies", *Mutation Research*, vol. 499, no. 1, pp. 13–25, 2002. doi: 10.1016/s0027-5107(01)00292-5. PMID: 11804602.

Index